Parametric Estimation

McGRAW-HILL SERIES IN PROBABILITY AND STATISTICS

David Blackwell and Herbert Solomon, Consulting Editors

Bharucha-Reid Elements of the Theory of Markov Processes and Their Applications
Drake Fundamentals of Applied Probability Theory
Dubins and Savage How to Gamble If You Must
Ehrenfeld and Littauer Introduction to Statistical Methods
Graybill Introduction to Linear Statistical Models, Volume I
Jeffrey The Logic of Decision
Li Introduction to Experimental Statistics
Miller Simultaneous Statistical Inference
Mood and Graybill Introduction to the Theory of Statistics
Morrison Multivariate Statistical Methods
Pearce Biological Statistics: An Introduction
Pfeiffer Concepts of Probability Theory
Raj Sampling Theory
Thomasian The Structure of Probability Theory with Applications
Wadsworth and Bryan Introduction to Probability and Random Variables
Wasan Parametric Estimation
Weiss Statistical Decision Theory
Wolf Elements of Probability and Statistics

PARAMETRIC ESTIMATION

M. T. WASAN
Professor of Mathematics
Queen's University
Kingston, Ontario

McGraw-Hill Book Company
New York, St. Louis, San Francisco, London
Sydney, Toronto, Mexico, Panama

Parametric Estimation

Copyright © 1970 by McGraw-Hill, Inc. All rights reserved. Printed in the United States of America. No part of this publication may be reproduced, stored in a retrieval system, or transmitted, in any form or by any means, electronic, mechanical, photocopying, recording, or otherwise, without the prior written permission of the publisher.

Library of Congress Catalog Card Number 74-85172

68400

1234567890MAMM7654321069

This book was set in Baskerville by The Maple Press Company, and printed on permanent paper and bound by The Maple Press Company. The designer was Merrill Haber. The editors were Donald K. Prentiss and Eva Marie Strock. Stuart Levine supervised the production.

To My Teachers

Preface

The subject of parametric estimation is a branch of statistical inference. There are many excellent books on inferential techniques, and some results of parametric estimation are mentioned in almost every one of them, but there does not exist one book devoted solely to this topic. In this volume we deal mostly with the case of univariate density and the following topics.

In the introduction we introduce various concepts of estimation with their anatomy so that one can see at a glance what a modern theory of estimation can provide as solutions to practical problems. In the second chapter we discuss the most ancient method of estimation, namely, the principle of least squares and its application to very recent modern problems of spacing from order statistics. Spacing for the regressor variables for least squares constitutes the third chapter.

The fourth chapter covers the concept of completeness, and the fifth chapter discusses the concept of sufficiency. These are the basic tools of modern statistics. Some properties of convex functions are appended in the sixth chapter, which will help in developing various methods of estimation.

In the seventh chapter we develop various methods of finding an unbiased estimate and append its properties by illustrative examples. The various concepts of unbiasedness are considered and their comparison with respect to various loss functions are investigated. In the eighth chapter different methods of finding lower bounds to the variance of an estimator are suggested.

The ninth chapter deals with the principle of maximum likelihood estimation and discusses its large sample theory, including its applications to problems of grouped samples and reliability.

In the tenth chapter the principles of admissibility and minimaxity are introduced. The various methods of finding a minimax estimator are given and are illustrated by examples. Their relation to game theory and to the principle of invariance is also appended.

The last chapter discusses the concept of empirical Bayes estimation, and a number of illustrative examples are given.

Almost all chapters are provided with various illustrative examples, and a number of problems are added at the end of each chapter to provide additional information and to help find out whether the reader has digested the material of the chapter. There are several very important methods of estimation not dealt with in this volume: principle of estimation by confidence sets, fiducial inference, the principle of invariance, large sample theory, sequential estimation, nonparametric estimation, methods of estimation of parameters of multivariate densities, techniques of estimation for stochastic processes, etc. A volume covering these topics is being planned for the near future.

This volume is a direct product of my lecture notes on the theory of parametric estimation given at Queen's University over a period of several years. I owe a great deal to successive groups of graduate students to whom I lectured, for various improvements in style and presentation and for the detection of errors of omission and commission. I extend my gratitude to Drs. A. K. Md. Ehsanes Saleh and J. R. Rutherford for their help in preparing Chapters 2 and 11, respectively.

I am especially indebted to Professor Z. Govindarajulu, who adopted this manuscript for class use at Case-Western Reserve University and provided me with an exhaustive list of errata and comments, and to

Professor S. Zacks, who used it for his course at Kansas State University and provided me with very helpful comments and suggestions.

I take pleasure in thanking Professors R. J. Buehler, H. A. David, J. Kiefer, E. L. Lehmann, H. Soloman, R. F. Tate, and many others for their comments and encouragement. Lastly, I would like to thank Mrs. E. M. Wight for her excellent typing of the manuscript.

As regards prerequisites for this volume, one needs a course in mathematical statistics at the level of Mood and Graybill, "Introduction to Theory of Statistics," 2d ed., McGraw-Hill, 1960, and a course in probability theory at the level of Parzen, "Modern Probability Theory and Its Applications," Wiley, 1960. Thus this book will be suitable for seniors or first-year graduate students.

It can be used for a two-semester course on estimation theory covering all the chapters and also for a one-semester course covering only part of the book, depending upon the interest of the instructor. For example, one can cover the materials of Chapters 4 to 10 in one semester or a quarter course. Since the proofs of the starred (*) sections assume a background in special branches of pure mathematics, it is advisable that they be omitted on the first reading.

Part of this work was done by the author as a Fellow of the Summer Research Institute of the Canadian Mathematical Congress.

M. T. WASAN

Contents

PREFACE   vii

ONE ☐ INTRODUCTION   1

    1. Introduction *1*
    2. The Principle of Point Estimation *2*
    3. Sufficient Statistics *3*
    4. Consistency *4*
    5. Estimation by a Method of Moments *5*
    6. The Least-squares Principle *6*
    7. Minimum-chi-square Method *7*
    8. Method of Maximum Likelihood *7*
    9. The Principle of Unbiasedness *8*
   10. Minimum-mean-square-error Estimate *9*
   11. Uniformly Minimum-variance-unbiased Estimate *9*
   12. The Principle of Minimaxity *10*
   13. The Principle of Invariance *11*

14. Interval Estimation *12*
15. Fiducial Probability *13*
16. The Large-sample Theory *14*
17. Sequential Estimation *14*
    References *15*

TWO ☐ LEAST-SQUARES ESTIMATION METHOD   18

1. Introduction *18*
2. Least-squares Estimation of Regression Coefficients *19*
3. The Estimation of Location and Scale Parameters of a Distribution *26*
4. An Illustration *29*
5. Determination of Optimum Quantiles for the Best Linear Unbiased Estimates of B when A = 0 *31*
   Problems *32*
   References *35*

THREE ☐ OPTIMAL SPACING OF A REGRESSOR VARIABLE   36

1. Introduction *36*
2. De la Garza Theorem *37*
3. The Optimal Location of the Regressor Variables *46*
4. Spacing for Minimax Variance *50*
   Problems *53*
   References *55*

FOUR ☐ COMPLETENESS   56

1. Introduction *56*
2. Complete Families of Probability Measures *57*
3. Boundedly Complete Families of Probability Measures *68*
4. Strongly Complete Families of Probability Measures *69*
   Problems *71*
   References *73*

FIVE ☐ SUFFICIENCY   74

1. Introduction *74*
2. Sufficient Statistics *75*

3. Families of Distributions Admitting Sufficient Statistics *78*
4. Sufficiency in the Dominated Case *81*
5. Minimal Sufficient Statistic *89*
   Problems *92*
   References *94*

SIX ☐ CONVEX FUNCTIONS 96

1. Introduction *96*
2. Examples of Convex Functions *97*
3. Sufficient Conditions for Convexity *97*
4. Properties of Convex Functions *98*
   Problems *102*
   References *102*

SEVEN ☐ UNBIASED ESTIMATION 104

1. Introduction *104*
2. Some Definitions *105*
3. Locally Best Unbiased Estimate *109*
4. The Use of a Complete, Sufficient Statistics to Construct Mean-unbiased Estimates *111*
5. Median Unbiasedness *118*
6. Modal Unbiasedness *120*
7. Relative Efficiency of the Modal-unbiased Estimate and Mean-unbiased Estimate *125*
   Problems *129*
   References *132*

EIGHT ☐ A LOWER BOUND OF THE VARIANCE OF AN ESTIMATE 134

1. Introduction *134*
2. The Cramer-Rao Inequality *135*
3. Cramer-Rao Lower Bound with Nuisance Parameters *139*
4. Chapman-Robbins-Kiefer Inequality *142*
5. Bhattacharyya Bounds *144*
   Problems *146*
   References *148*

NINE ☐ MAXIMUM-LIKELIHOOD ESTIMATION 150

1. Introduction *150*
2. The Method Illustrated *151*

3. Large-sample-size Properties *157*
4. Numerical Solutions of Maximum-likelihood Equations *161*
5. On Reliability Inference *162*
   Problems *172*
   References *178*

TEN □ ADMISSIBLE AND MINIMAX ESTIMATION 182

1. Introduction *182*
2. Admissible Estimate *183*
3. Theorems and Their Applications *184*
4. Binomial Minimax Estimation for a Given Stopping Rule *191*
5. Minimax Estimates by the Cramer-Rao Inequality Method *197*
6. Bayes Estimate for Quadratic Loss Function *202*
7. Bayes Estimates where the Loss Depends on the Absolute Error *206*
8. Relation of the Principles of Invariance, Minimaxity, and Admissibility *207*
   Problems *212*
   References *216*

ELEVEN □ THE EMPIRICAL BAYES METHOD OF ESTIMATION 218

1. Introduction *218*
2. An Empirical Bayes Technique *219*
3. Optimality Criterion *226*
4. A Case in Which Asymptotic Optimality Does Not Exist but All is Not Lost *231*
   Smooth Empirical Bayes Estimation *234*
   Problems *238*
   References *242*

INDEX **247**

Parametric Estimation

chapter One

Introduction

1 □ INTRODUCTION

The analysis of real and complex variables has influenced not only physics, engineering, and many other sciences but also many other branches of mathematics. Approximation theory particularly is its main beneficiary. One of the twentieth-century contributions to mathematics is in the field of analysis of random variables (probability and stochastic processes). This has now contributed to almost every scientific discipline and especially to mathematical statistics, whose development went hand in hand with the development of probability theory (sometimes it is even hard to distinguish between them). Its main beneficiary is the theory of estimation.

Human nature is always investigative, but the investigations of the last couple of decades are remarkable: they have brought a revolution in space and along with it revolution in every branch of science. Experimentation has provided a stimulus to the imagination, but this has not been enough; thus it led to a science, called *statistics*, where mathematical inferential techniques are developed. *Estimation theory* is a set of some of those techniques.

The author's experience in dealing with applications of the theory in industrial problems has been that an experimenter should diagnose the parent population, its form, and its special characteristics and set up a set of guidelines dictated by practical situations in drawing samples and using inferential techniques. As the word *estimation* indicates, inferential technique leads to an approximation of the real situation, which is unknown. During the Second World War the need for these techniques was great. Of course, in subsequent years it grew even greater, as is apparent in the voluminous applied and theoretical literature which has appeared on the subject. It was Wald who formalized these situations as problems of decision theory. We now give his formalization.

Let D be a space of decisions. Let d be any member of this class. We are in an unknown situation. The true situation cannot be determined, either because it is too expensive or impossible. But we have a random variable X (real, vector, or sequence) whose distribution function depends on the true situation. The observed value x of X provides information about the true situation, which can be used to increase our chances of making a good decision. Then the problem is to choose a decision function $\delta(x)$ from the sample space to the decision space. Let $X = x$ and $\delta(x) \in D$; then take decision $\delta(x)$. If it is a desirable one, it usually results in a good decision.

2 □ THE PRINCIPLE OF POINT ESTIMATION

Let us make a decision $d \in D$ and suppose that X has a distribution function F in the true situation. Then let $L(F,d)$ be the loss. $L(F,d)$ is a real-valued nonnegative function of F and d. The desirability of any particular decision rule $\delta(X)$ is determined by its risk function

$$R_\delta(F) = E_F\{L[F,\delta(X)]\} = \int_{-\infty}^{\infty} L[F,\delta(x)]\, dF(x)$$

Let us now assume that the form of F is known but depends on an unknown parameter. Thus we are interested in finding an estimate of this unknown parameter. In this case, $\delta(X)$ is a point estimate of the parameter which is not known. We illustrate this by the following example.

EXAMPLE

Let X have distribution function $F_\lambda(x)$, and let λ be the true situation or real parameter which we want to determine. Let $[\delta(X) - \lambda]^2$ be the loss function.

$$R_\delta(\lambda) = E_\lambda[\delta(X) - \lambda]^2 = \int_{-\infty}^{\infty} [\delta(x) - \lambda]^2 \, dF_\lambda(x)$$

If

$$R_\delta(\lambda) \leq E_{\delta*}(\lambda) \qquad \text{for all } \lambda$$

whatever the δ^*, then $\delta(X)$ is a uniformly best estimate of λ. Now we have to consider: Does it exist? If it does, then what is that function of X?

We turn now to the language of estimation theory.

3 □ SUFFICIENT STATISTICS

The concept of *sufficiency* was introduced by Fisher. It is a very important concept from the point of view of applications and for developing inferential techniques which may be used to analyze the data of an experiment. Usually an experimenter has a number of observations, and from just looking at them it is generally not possible to draw an inference. Thus the experimenter is interested in constructing a function of the observations, and this function should yield all he wants to know. For example, if he is interested in finding an estimate of an unknown parameter, he should be able to construct a *reliable* estimate.

Let $X_1, \ldots, X_n$ be independent random variables with density function $f_\lambda(x)$. Let $T(X_1 \cdots X_n)$ be a function of $X_1 \cdots X_n$ such that the conditional distribution function of X_i ($i = 1 \cdots n$), given T, is independent of λ or it does not depend on λ, then $T(X_1 \cdots X_n)$ is said to be a *sufficient statistic*.

Complete family of density functions

Let X be a random variable with density function $f_\lambda(x)$ such that

$$\int_{-\infty}^{\infty} \phi(x) f_\lambda(x) \, dx = 0 \Rightarrow \phi(x) = 0 \qquad \text{almost everywhere (a.e.) for every } \lambda$$

Then $f_\lambda(x)$ is said to be a complete family of density functions. This concept plays an important role in establishing the uniqueness of an unbiased estimate, etc.

4 □ CONSISTENCY

Let $X_1, X_2, \ldots, X_n, \ldots$ be independent random variables with density function $f_\lambda(x)$ depending on an unknown parameter $\lambda \in \Lambda$. Let T_n be a function of $X_1, X_2, \ldots, X_n$. Then $\{T_n\}$ is said to be a *consistent sequence* of estimates for θ if given $\delta > 0$, there exists $\varepsilon > 0$ and N such that
$$P\{|T_n - \lambda| > \delta\} < \varepsilon \qquad \text{for } n > N$$

Definition

In short, if $T_n \to \lambda$ in probability as $n \to \infty$, T_n is named a consistent estimate of λ.

EXAMPLE 1
Let $X_1, X_2, \ldots, X_n, \ldots$ be independent random variables, each normally distributed with mean λ and known variance σ^2; then $\bar{X}_n$ is a consistent estimate of λ. For given $\delta > 0$, by Tchebychev's inequality
$$P_r\{|\bar{X}_n - \lambda| > \delta\} < \frac{\sigma^2}{n\delta^2}$$

Hence $\{\bar{X}_n\}$ is consistent sequence of estimates for λ.

EXAMPLE 2
Let $X_1, X_2, \ldots, X_n \ldots$ be independent random variables with common density function
$$f_\lambda(x) = \frac{1}{\pi[1 + (x - \lambda)^2]} \qquad \begin{matrix} -\infty < \lambda < \infty \\ -\infty < x < \infty \end{matrix}$$

Then the density function of $\bar{X}_n$ is same as that of a single variable. Hence for $\delta > 0$
$$P\{|\bar{X}_n - \lambda| > \delta\}$$
remains the same for every n; hence $\bar{X}_n$ is not a consistent estimate of λ.

Thus the basic problem here is what the necessary and sufficient conditions are for the existence of a consistent sequence of estimates for an unknown parameter λ. LeCam and his colleagues [17, 18] have investigated this problem.

5 □ ESTIMATION BY A METHOD OF MOMENTS

Let $X_1, X_2, \ldots, X_n$ be independent random variables with a common density function depending on the unknown parameters $\lambda_1, \lambda_2, \ldots, \lambda_k$. Let us assume that the first k moments about the origin of the distribution function exist as functions $\phi_r(\lambda_1, \ldots, \lambda_k)$, where $r = 1, 2, \ldots, k$, of the parameters. Let us denote by

$$\hat{\phi}_r = \frac{1}{n} \sum_1^n X_i^r \qquad (1)$$

the sample rth moment about the origin. Then by equating $\hat{\phi}_r$ to ϕ_r, where $r = 1, \ldots, k$, we have k equations in k unknown parameters. Thus one can get an estimate of a parameter. This method is known as the *method of moments*. Under suitable conditions this method generates a consistent sequence of estimates of a parameter. It is one of the oldest methods and played an important role in determining the Pearson type of distributions. An example will demonstrate the technique.

EXAMPLE

Let $X_1, X_2, \ldots, X_n$ be independent random variables with the density function

$$f(x, \lambda_1, \lambda_2) = \frac{1}{\sqrt{2\pi \lambda_2}} e^{-(x-\lambda_1)^2/2\lambda_2} \qquad \begin{array}{c} -\infty < x < \infty \\ -\infty < \lambda_1 < \infty \\ \lambda_2 > 0 \end{array} \qquad (2)$$

Then it is well known that

$$E(X) = \lambda_1 \qquad (3)$$
$$E(X^2) = \lambda_2 + \lambda_1^2 \qquad (4)$$

Therefore

$$\hat{\lambda}_1 = \frac{1}{n} \sum_1^n X_i = \bar{X} \qquad (5)$$

6 PARAMETRIC ESTIMATION

$$\hat{\lambda}_2 = \frac{1}{n} \sum X_i^2 - \left(\frac{1}{n} \sum X_i\right)^2$$
$$= \frac{1}{n} \left[\sum_{1}^{n} (X_i - \bar{X})^2\right] \tag{6}$$

For further details of the method, see Fisher [8], who discusses the techniques.

6 □ THE LEAST - SQUARES PRINCIPLE

The method of least squares goes back to Gauss, who first developed this technique of estimation. Generally it is used for the estimation of parameters in a linear model. We shall discuss this method briefly.

Let $X_1, X_2, \ldots, X_n$ be independent random variables such that $E(X_i) = \lambda_1 + \lambda_2 y_i$, where y_i is a known constant and λ_1 and λ_2 are unknown constants. Let us consider the quadratic function

$$W(X_1, \ldots, X_n, \lambda_1, \lambda_2) = \sum_{i=1}^{n} (X_i - \lambda_1 - \lambda_2 y_i)^2 \tag{1}$$

Let $\hat{\lambda}_1$ and $\hat{\lambda}_2$ be the estimates of λ_1 and λ_2, respectively, which minimize (1); then $\hat{\lambda}_1$ and $\hat{\lambda}_2$ are called *least-squares estimates* of λ_1 and λ_2. And $\hat{\lambda}_1$ and $\hat{\lambda}_2$ can be obtained by solving the following so-called normal equations:

$$\frac{\partial W(X_1, \ldots, X_n, \lambda_1, \lambda_2)}{\partial \lambda_1} = -2 \sum_{i=1}^{n} (X_i - \lambda_1 - \lambda_2 y_i) = 0$$

or

$$\sum_{1}^{n} X_i = n\lambda_1 + \lambda_2 \sum y_i \tag{2}$$

and

$$\frac{\partial W(X_1, \ldots, X_n, \lambda_1, \lambda_2)}{\partial \lambda_2} = -2 \sum_{i=1}^{n} y_i(X_i - \lambda_1 - \lambda_2 y_i) = 0$$

or

$$\sum_{1}^{n} y_i X_i = \lambda_1 \sum y_i + \lambda_2 \sum_{1}^{n} y_i^2 \tag{3}$$

By solving (2) and (3) one can obtain

$$\hat{\lambda}_2 = \frac{\sum_{i=1}^{n} (X_i - \bar{X})(Y_i - \bar{Y})}{\sum_{1}^{n} (Y_i - \bar{Y})^2}$$

and

$$\hat{\lambda}_1 = \bar{X} - \hat{\lambda}_2 \bar{Y}$$

We shall discuss this method in Chaps. 2 and 3. We shall also discuss various problems one can solve with this technique. The large number of papers and books which have appeared on the subject speak for its usefulness.

7 □ MINIMUM - CHI - SQUARE METHOD

This method is applicable where the observed values of random variables themselves are frequencies of a finite number of mutually exclusive events with probabilities $p_1, p_2, \ldots, p_m$, which are functions of k unknown parameters λ to be estimated. Our data consist of $n_1, n_2, \ldots, n_m$ observed frequencies, respectively, for the m mutually exclusive events. Let $n = \sum_{1}^{m} n_i$, and obviously $np_1, \ldots, np_m$ are the expected frequencies of the events. Then one can use as the measure of discrepancy

$$\chi^2 = \sum_{1}^{m} \frac{(x_i - np_i)^2}{np_i} \quad (1)$$

where p_i's are functions of λ. Then $\hat{\lambda}$ is called the *minimum-chi-square estimate* of λ if it minimizes (1). This estimate has very nice asymptotic properties, which are discussed by Neyman [22].

8 □ METHOD OF MAXIMUM LIKELIHOOD

Let $X_1, X_2, \ldots, X_n$ be the random variables with density function

8 PARAMETRIC ESTIMATION

$f(x,\lambda_1, \ldots ,\lambda_k)$. Then the likelihood function is defined as

$$L(X_1, \ldots ,X_n;\lambda_1, \ldots ,\lambda_k) = \prod_{i=1}^{n} f(x_i;\lambda_1, \ldots ,\lambda_k) \qquad (1)$$

$(\hat{\lambda}_1, \ldots ,\hat{\lambda}_k)$ is said to be the maximum-likelihood estimate of $(\lambda_1, \ldots ,\lambda_k)$ if it maximizes the likelihood function (1). We assume the existence of the estimates and the maximum of (1). This concept was introduced by Fisher [9], who investigated a number of its properties and its relation to other methods of estimation. We now give an illustrative example.

EXAMPLE

Let $X_1, X_2, \ldots , X_n$ be random variables which are normally distributed with mean λ_1 and variance λ_2. Then

$$L(x_1, \ldots ,x_n;\lambda_1,\lambda_2) = \prod_{i=1}^{n} \frac{1}{\sqrt{2\pi\lambda_2}} e^{-(x_i-\lambda_1)^2/2\lambda_2} \quad \begin{array}{c} -\infty < x_i < \infty \\ -\infty < \lambda_1 < \infty \\ \lambda_2 > 0 \end{array} \qquad (2)$$

$$\log L(x_1, \ldots ,x_n;\lambda_1,\lambda_2) = -\frac{n}{2} \log \pi - \frac{n}{2} \log \lambda_2 - \frac{n}{2} \log 2 - \frac{\sum_{1}^{n}(x_i-\lambda_1)^2}{2\lambda_2} \qquad (3)$$

$$\frac{\partial \log L}{\partial \lambda_1} = -\lambda \sum_{1}^{n} \frac{x_i-\lambda_1}{\lambda_2} = 0 \qquad (4)$$

$$\frac{\partial \log L}{\partial \lambda_2} = -\frac{n}{2}\frac{1}{\lambda_2} + \sum_{1}^{n} \frac{(x_i-\lambda_1)^2}{2\lambda_2^2} = 0 \qquad (5)$$

By solving (4) and (5) one can obtain

$$\hat{\lambda}_1 = \bar{x} = \frac{1}{n}\sum_{1}^{n} x_i \qquad (6)$$

$$\hat{\lambda}_2 = \frac{1}{n}\sum_{1}^{n} (x_i-\bar{x})^2 \qquad (7)$$

9 □ THE PRINCIPLE OF UNBIASEDNESS

Let $X_1, X_2, \ldots , X_n$ be independent random variables with density function $f(x,\lambda)$. Then $\phi(X_1, \ldots , X_n)$, a function of $X_1, \ldots ,X_n$, is said to be a *mean-unbiased estimate* of λ if

$$E[\phi(X_1, \ldots, X_n)] = \lambda \qquad (1)$$

If

$$E[\phi_1(X_1, \ldots, X_n)] = \lambda + b \qquad (2)$$

$\phi_1(X_1, \ldots, X_n)$ is said to be a *biased estimate* of λ with expected bias b. The principle of mean unbiasedness shows that it is a fair principle because it favours all values of λ impartially.

10 □ MINIMUM - MEAN - SQUARE - ERROR ESTIMATE

Let us consider $L(\phi,\lambda) = (\phi - \lambda)^2$ as a loss function, as is done for least-squares estimation. We do not know the value of λ. We estimate it by ϕ; if a value of ϕ is close enough to λ, the corresponding loss will be small; if it is not, then the loss will be large. Thus we shall be interested in finding ϕ such that

$$E(\phi - \lambda)^2 \leq E(\phi' - \lambda)^2 \qquad \text{for every } \phi' \qquad (1)$$

Then ϕ has minimum mean square error, and we call it a *minimum-mean-square-error estimate*.

Minimum - variance - unbiased estimate

Now, if in addition, we assume that ϕ and ϕ' are unbiased estimates of λ, then (1) implies that ϕ has minimum variance, and hence it is a minimum-variance-unbiased estimate.

11 □ UNIFORMLY MINIMUM VARIANCE - UNBIASED ESTIMATE

Now let ϕ and ϕ' be unbiased and $E(\phi - \lambda)^2 \leq E(\phi' - \lambda)^2$ for every ϕ unbiased and every λ. Then ϕ is said to be a uniformly minimum-variance-unbiased estimate.

EXAMPLE 1

Let $X_1, X_2, \ldots, X_n$ be random variables normally distributed with mean λ_1 and variance λ_2. Then

$$\frac{\sum_{i=1}^{n} (X_i - \bar{X})^2}{n - 1}$$

is an unbiased estimate of λ_2. But one can easily verify that it is not a minimum-mean-square-error estimate. In fact one can show that

$$E\left[\frac{\sum_{1}^{n} (X_i - \bar{X})^2}{n + 1} - \lambda_2\right]^2 < E\left[\frac{\sum_{1}^{n} (X_i - \bar{X})^2}{n - 1} - \lambda_2\right]^2$$

Thus the choice of estimate depends on the criterion preferred.

EXAMPLE 2

Let us assume the hypothesis of Example 1. Then $\bar{X}$ is an unbiased estimate of λ_1; not only that, it is a uniformly minimum-variance-unbiased estimate.

REMARK

There are a number of ways one can define unbiasedness. We shall discuss them and their relationships in the chapter on unbiased estimation.

12 □ THE PRINCIPLE OF MINIMAXITY

Von Neumann developed the theory of games and introduced minimax and maximin principles. At the same time, Wald used the minimax principle in the construction of minimax estimates.

Let $X_1, \ldots, X_n$ be independent random variables with density function $f_\lambda(x)$, where λ is unknown but belongs to the set Λ. Let $L(\delta,\lambda)$ be the loss function. Then $\delta(X_1, \ldots, X_n)$ is said to be *minimax estimate* of λ if

$$\sup_{\lambda \in \Lambda} E\{L[\delta(X_1, \ldots, X_n),\lambda]\} \leq \sup_{\lambda \in \Lambda} E\{L[\delta^*(X_1, \ldots, X_n),\lambda]\}$$

for every δ^*

Thus the principle is used to guard against the worst situation that can turn up. Generally, in the construction of this estimate, one uses an a priori distribution of an unknown parameter, which is a lot to assume. But it is quite reliable for a small-sample estimation. Closely related to this principle is the concept of *admissibility*, discussed in Chap. 10, where we also discuss the relation of these concepts to the principle of invariance.

13 □ THE PRINCIPLE OF INVARIANCE

Let X be a random variable defined on $\mathfrak{X}$. Let $\mathfrak{F}$ be a Borel field of the subsets of $\mathfrak{X}$, and let P_x be a probability measure. Suppose that one is interested in finding an estimate of λ; if one knows that there exists a transformation g of x such that it maps $\mathfrak{X}$ into $\mathfrak{X}$ and the probability measure induced by g is the same as it was before, then the mapping g is called *invariant*, because it leaves the probability structure of the problem unchanged. Thus one can apply the transformation to the observed values of the random variables and can estimate the transformed value of the parameter λ. This method will give exactly the same solution that one was interested in finding in the first place.

EXAMPLE

Let X be a random variable with density functions $f(x,\lambda)$, where λ is a location parameter. Let $\delta(X)$ be an estimate of λ and

$$R(\lambda) = E[\delta(X) - \lambda]^2$$

the risk of the estimate of λ. Then one should be willing to estimate $\lambda + c$, where c is known constant, by $\delta(X) + c$. Because the new problem is reduced to the original one, the risk remains invariant also. Thus any estimate δ for which $\delta(X + c) = \delta(X) + c$ is called an *invariant estimate* of λ. In general we consider a transformation on X and a corresponding transformation on λ and also their effect on the risk or expected loss function.

We state the following results due to Hunt, which are quite useful; the proofs of these results are simple.

(i) Let g be a measurable one-to-one transformation on $\mathfrak{X}$ to $\mathfrak{X}$, and let $\bar{g}$ be the transformation induced on the parametric space Λ to Λ. Then $\bar{g}$ is a one-to-one transformation Λ to Λ.

(ii) Let $\mathcal{G}$ be an arbitrary set of elements g, with operations of product and inverse defined. Assume

$$g \in \mathcal{G} \Rightarrow g \text{ has some property } s \tag{1}$$
$$g_1 \text{ has property } s \text{ and } g_2 \text{ has property } s \Rightarrow g_1 g_2$$
$$\text{has property } s \tag{2}$$
$$g \text{ has property } s \Rightarrow g^{-1} \text{ has property } s \tag{3}$$

Then all elements of the group G, generated by $\mathcal{G}$, have the property s.

(iii) Let $\mathcal{G}$ be a set of transformations on $\mathcal{X}$ to $\mathcal{X}$ such that for $g \in \mathcal{G}$,

$$A \in \mathcal{F} \Rightarrow gA \in \mathcal{F} \text{ and } g^{-1}A \in \mathcal{F} \tag{4}$$
$$\bar{o}r = r \quad \text{and} \quad \bar{o}(\Lambda - r) = \Lambda - r \quad \text{where } r \subset \Lambda \tag{5}$$

Then the transformations of the group G, generated by $\mathcal{G}$, have properties (4) and (5) of result (iii).

(iv) Assuming the setup of (iii), the set $\bar{G}$ of all $\bar{g}$ corresponding to the transformations $g \in G$ is a group which is homomorphic to G.

(v) Let δ be an invariant estimate of λ and L be an invariant loss function; then

$$R_\delta(\lambda) = R_\delta(\bar{g}\lambda) \qquad \text{for all } g \in \mathcal{G}$$
$$\lambda \in \Lambda$$

where R is the expected loss.

So far we have discussed one point-estimation method; now we define and illustrate estimation by the confidence-set method.

14 □ INTERVAL ESTIMATION

The concept of an interval estimation was first defined by Neyman [21], who showed its relation to problems of testing hypotheses.

Let $X_1, X_2, \ldots, X_n$ be independent random variables with a common density function $f_\lambda(x)$. Let $\phi_\lambda(X_1, \ldots, X_n)$ be a function of $X_1, \ldots, X_n$ and λ such that the density function g of ϕ does not involve λ. Then one can find $t_1 < t_2$ such that

$$\int_{t_1}^{t_2} g(\phi) \, d\phi = 1 - \alpha \qquad \text{where } 0 < \alpha < 1$$

Assuming that the inverse of ϕ exists, one can find, for given values of $X_1, \ldots, X_n$, lower and upper bounds of λ as functions of $X_1, \ldots, X_n$ and α, say $[\underline{\lambda}, \bar{\lambda}]$, which is called a $1 - \alpha$ confidence interval for λ. We shall demonstrate this method of estimation by the following well-known example.

EXAMPLE
Let $X_1, \ldots, X_n$ be random variables normally distributed with mean λ and variance σ^2, both unknown. It is well known that

$$\frac{\bar{X} - \lambda}{\sqrt{\Sigma(X_i - \bar{X})^2/n(n-1)}}$$

is distributed as t with $n - 1$ degrees of freedom and the distribution does not depend on λ.
Let

$$\Pr\left[\left|\frac{\bar{X} - \lambda}{\sqrt{\Sigma(X_i - X)^2/n(n-1)}}\right| < t_\alpha\right] = 1 - \alpha$$

$$= \Pr\left[\bar{X} - t_\alpha \sqrt{\frac{\Sigma(X_i - \bar{X})^2}{n(n-1)}} \le \lambda \le \bar{X} + t_\alpha \sqrt{\frac{\Sigma(X_i - \bar{X})^2}{n(n-1)}}\right] = 1 - \alpha$$

Hence

$$\left[\bar{X} - t_\alpha \sqrt{\frac{\Sigma(X_i - \bar{X})^2}{n(n-1)}},\ \bar{X} + t_\alpha \sqrt{\frac{\Sigma(X_i - \bar{X})^2}{n(n-1)}}\right]$$

is a $1 - \alpha$ confidence interval for λ or an interval estimate of λ.

15 □ FIDUCIAL PROBABILITY

The concept of fiducial probability was developed by Fisher. We illustrate by the previous example but assume that $\sigma^2 = 1$.

$$P\left(\frac{\bar{X} - \lambda}{\sqrt{1/n}} < z\right) = \int_{-\infty}^{z} \frac{1}{\sqrt{2\pi}} e^{-v^2/2}\, dy = \Phi(z)$$

14   PARAMETRIC ESTIMATION

Let $\bar{X}$ take the value 2.5; then the fiducial argument yields that

$$P\left(2.5 - \lambda < \frac{z}{\sqrt{n}}\right) = P\left(\lambda > 2.5 - \frac{z}{\sqrt{n}}\right) = \bar{\Phi}(z) \qquad (1)$$

Several objections are raised to this statement. First, λ may not be a random variable; second, even if it is a random variable with an unknown a priori distribution, then statement (1) depends on it and needs a different interpretation, etc. The work of Dempster [6], Fraser [11], and Sprott [27] has supplied answers to some of these objections.

16 □ THE LARGE - SAMPLE THEORY

Let $X_1, X_2, \ldots, X_n, \ldots$ be independent random variables with the same density function $f_\lambda(x)$. One is interested in finding an estimate of λ when n, the number of observations, is very large. Then the estimation procedure for λ falls in the domain of large-sample theory. We have already discussed the concept of consistency, which is a concept of large-sample theory. Another concept is the "best asymptotically normal estimate," which was developed by Neyman [21] from the minimum-chi-square method. We shall see some large-sample properties of maximum-likelihood estimates in Chap. 9, which will provide illustrative examples for this theory. When the random variables are dependent, various large-sample properties of the maximum-likelihood estimates of a parameter are discussed by Wald.

Another very important concept, the empirical Bayes approach to problems of estimation and decision in general, has been developed by Robbins [25]; we shall discuss it in the last chapter.

17 □ SEQUENTIAL ESTIMATION

In the techniques discussed so far, we have assumed that we have a fixed number n of observations of a random variable having a certain density function. In other words, the sample space $(\mathfrak{X}, \mathfrak{F}, P_\lambda)$ is fixed. Now we obtain the observations sequentially; i.e., we obtain one after another and make an estimate of λ, every time making use of the information available so far. Such a procedure might reduce the number of observations, which are usually costly, and thus reduce the cost of experimentation. During the Second World War Wald and Barnard independently developed techniques of sequential analysis for problems of testing hypotheses. After the war the sequential-estimation problem received increased attention.

Wald [29] extended his theory of minimaxity to the sequential case, and Wolfowitz [32] solved problems of unbiased sequential estimation. Basically a sequential-estimation problem is a hard one, because with each observation the sample space changes and one cannot bypass a distribution problem for estimation, as can be done for the testing-hypothesis situation.

In 1946 Girshick, Mosteller, and Savage [15] considered a simple binomial sampling, where an observation can take the value 0 or 1. Since one can write the distribution for the given stopping rule explicitly, estimation can be handled easily. DeGroot [5] showed optimality of the fixed-sample procedure, and Wasan [30] showed optimality of the curtailed symmetric rule for unbiased estimation of a parameter.

Robbins [24] developed a sequential-estimation procedure for the mean of the normal density. The area of sequential estimation is under active investigation, but progress is very slow, due to the difficulty pointed out in the second paragraph of this section.

The concept of sequential analysis has influenced many other areas of mathematics, namely, the field of programming (see Bellman [1]) and the field of stochastic approximation, i.e., stochastic numerical analysis (see Wasan [31]).

REFERENCES

1 BELLMAN, R.: "Dynamic Programming," Princeton University Press, Princeton, N.J., 1957.

2 BLACKWELL, D., and M. A. GIRSHICK: "Theory of Games and Statistical Decision Functions," John Wiley & Sons, Inc., New York, 1954.

3 CHERNOFF, H.: Large Sample Theory Parametric Case, *Ann. Math. Statist.*, vol. 21, pp. 1–22, 1956.

4 CRAMÉR, H.: "Mathematical Methods of Statistics," Princeton University Press, Princeton, N.J., 1946.

5 DEGROOT, M. H.: Unbiased Binomial Sequential Estimation, *Ann. Math. Statist.*, vol. 30, pp. 80–101, 1959.

6 DEMPSTER, A. P.: On the Difficulties Inherent in Fisher's Fiducial Argument, *Proc. Intern. Statist. Conf.*, *Ottawa*, 1963.

7 DEUTCH, R.: "Estimation Theory," Prentice-Hall, Inc., Englewood Cliffs, N.J., 1965.

8 FISHER, R. A.: On the Mathematical Foundation of Theoretical Statistics, *Phil. Trans. Roy. Soc. London, ser. A*, vol. 222, pp. 309–368, 1922.

9 FISHER, R. A.: "Contributions to Mathematical Statistics," John Wiley & Sons, Inc., New York, 1950.

10 FISHER, R. A.: Two New Properties of Mathematical Likelihood, *Proc. Roy. Soc. London, ser. A*, vol. 144, p. 285, 1934.

11 FRASER, D. A. S.: The Fiducial Method and Invariance, *Biometrika*, vol. 48, pp. 261–280, 1961.

12 FRASER, D. A. S.: "Nonparametric Methods in Statistics," John Wiley & Sons, Inc., New York, 1957.

13 GAUSS, C. F.: Aus der Theorie der Bewegung der Himmels-Korper, *Abhandl. Methode Kleinsten Quadrate*, 1887.

14 GAUSS, C. F.: Theorie der den kleinsten Fehlern interworfenen Combination der Beobachtangen, *Abhandl. Methode Kleinsten Quadrate*, 1887.

15 GIRSHICK, M. A., F. MOSTELLER, and L. J. SAVAGE: Unbiased Estimates for Certain Binomial Sampling Problems with Applications, *Ann. Math. Statist.*, vol. 17, pp. 13–23, 1946.

16 KENDALL, M. G., and A. STUART: "Advanced Theory of Statistics," vols. 1 and 2, Charles Griffin & Company, Ltd., London, 1958.

17 LECAM, L., and L. SCHWARTZ: A Necessary and Sufficient Condition for the Existence of Consistent Estimates, *Ann. Math. Statist.*, vol. 37, pp. 140–150, 1960.

18 LECAM, L., L. BREIMAN, L. SCHWARTZ: Consistent Estimates and Zero-one Sets, *Ann. Math. Statist.*, vol. 35, pp. 157–161, 1964.

19 LEHMANN, E. L.: Notes on Theory of Estimation, University of California Press, Berkeley, Calif., 1950.

20 LEHMANN, E. L., and H. SCHEFFÉ: Completeness, Similar Regions and Unbiased Estimation, *Sankhya, ser. A*, vol. 10, pp. 305–340, 1950.

21 NEYMAN, J.: Outline of a Theory of Statistical Estimation Based on Classical Theory of Probability, *Phil. Trans. Royal Soc. London, ser. A.*, vol. 236, pp. 333–380, 1937.

22 NEYMAN, J.: Contribution to the Theory of the χ^2 Test, *Proc. 1st Berkeley Symp. Math. Statist. Probability*, pp. 239–273, 1949.

23 RAO, C. R.: "Linear Statistical Inference and Its Application," John Wiley & Sons, Inc., New York, 1965.

24 ROBBINS, H.: Sequential Estimation of the Mean of a Normal Population, in "Probability and Statistics" (Harold Cramér volume), pp. 235–245, Almquist and Wiksell, Uppsala, 1959.

25 ROBBINS, H.: The Empirical Bayes Approach to Statistical Decision Problems, *Ann. Math. Statist.*, vol. 35, pp. 1–20, 1964.

26 SCHEFFÉ, H.: "Analysis of Variance," John Wiley & Sons, Inc., New York, 1958.

27 Sprott, D. A.: Similarities between Likelihoods and Associated Distributions Posteriori, *J. Roy. Statist. Soc.*, ser. *B*, vol. 24, pp. 460–462, 1961.

28 Wald, A.: "Theory of Decision Functions," John Wiley & Sons, Inc., New York, 1950.

29 Wald, A.: Statistical Decision Functions Which Minimize the Maximum Risk, *Ann. Math. Statist.*, vol. 46, pp. 265–280, 1945.

30 Wasan, M. T.: Sequential Optimum Procedures for Unbiased Estimation of a Binomial Parameter, *Technometrics*, vol. 6, pp. 259–272, 1964.

31 Wasan, M. T.: "Stochastic Approximation," Cambridge University Press, New York, 1969.

32 Wolfowitz, J.: Efficiency of Sequential Estimates, *Ann. Math. Statist.*, vol. 18, pp. 215–230, 1947.

chapter Two

Least-squares Estimation Method

1 □ INTRODUCTION

In this chapter we discuss the application of the principle of least squares to problems of estimation of polynomial regression coefficients. We discuss the Gauss-Markoff theorem and give illustrative examples. We shall follow Scheffé [6] for the method of proof; further discussion will be found in this excellent text.

Least squares is one of the oldest methods of estimation, but we shall apply it to modern problems of estimation of location and scale parameters of a distribution and optimal spacing from order statistics. In Chap. 3 we apply this principle of least squares to the modern problem of optimal spacing of a regressor variable.

2 □ LEAST - SQUARES ESTIMATION OF REGRESSION COEFFICIENTS

Let us consider the polynomial regression model

$$Y_i = \gamma_0 + \gamma_1 x_i + \cdots + \gamma_k x_i^k + \varepsilon_i \qquad i = 1, \ldots, m \qquad (1)$$
$$m > k + 1$$

The Y_i and ε_i are random variables, and the x_i are the observed values of the regressor variable X which is a real variable. In matrix notation (1) becomes

$$\mathbf{Y} = X\boldsymbol{\gamma} + \boldsymbol{\varepsilon} \qquad (2)$$

where

$$\mathbf{Y}' = (Y_1, \ldots, Y_m) \qquad \boldsymbol{\varepsilon}' = (\varepsilon_1, \ldots, \varepsilon_m) \qquad \boldsymbol{\gamma}' = (\gamma_0, \ldots, \gamma_k)$$

and

$$X = \begin{bmatrix} 1 & x_1 & x_1^2 & \cdots & x_1^k \\ 1 & x_2 & x_2^2 & \cdots & x_2^k \\ \cdots & \cdots & \cdots & \cdots & \cdots \\ 1 & x_m & x_m^2 & \cdots & x_m^k \end{bmatrix}$$

The prime on a vector or matrix denotes the transpose. The random variable Y_i has the value y_i when $X = x_i$. Let us assume that $E(\mathbf{Y}) = X\boldsymbol{\gamma}$, that the rank of X is $k + 1$, and that the covariance matrix of the vector variable Y is given by $\Sigma_Y = \sigma^2 I$.

Let $\mathbf{g}' = (g_0, \ldots, g_k)$ be the estimate of $\boldsymbol{\gamma}' = (\gamma_0, \ldots, \gamma_k)$; then the squared error is

$$\boldsymbol{\varepsilon}'\boldsymbol{\varepsilon} = (\mathbf{Y} - X\mathbf{g})'(\mathbf{Y} - X\mathbf{g}) = |\mathbf{Y} - X\mathbf{g}|^2 \qquad (3)$$

The least-squares estimators, denoted by $\hat{\boldsymbol{\gamma}}$, are those functions of $\mathbf{Y}$ such that if $g_j = \hat{\gamma}_j$, $j = 0, \ldots, k$, then the error sum of squares (3) will be minimized. To obtain this minimum we differentiate (3) with respect to each g_j, set equal to zero, and solve the resulting equations for the g_j. Thus

$$\frac{\partial}{\partial g_j}(\mathbf{Y} - X\mathbf{g})'(\mathbf{Y} - X\mathbf{g}) = 0 \qquad j = 0, \ldots, k$$

reduces to the normal equations

$$(X'X)\mathbf{g} = X'\mathbf{Y} \qquad (4)$$

The inverse $(X'X)^{-1}$ exists since the rank of X is assumed to be $k + 1$.

Thus the least-squares estimators are given by

$$\hat{\gamma} = (X'X)^{-1}X'Y \tag{5}$$

Lemma 1

An estimator is called a least-squares estimate if and only if it satisfies the set of normal equations.

Proof:

Let the vector of random variables $\mathbf{Y}$ generate an m-dimensional vector space U_m. In this space let the vector of means be $\mathbf{\mu} = E(\mathbf{Y}) = X\gamma$. Thus

$$\mathbf{\mu} = \gamma_0 \mathbf{X}_0 + \cdots + \gamma_k \mathbf{X}_k$$

where X_j is the $(j+1)$st column of the matrix X, $j = 0, \ldots, k$. Let $s \leq k+1$ be the rank of X. Then an arbitrary vector $\mathbf{V}$ lies in U_s, the space spanned by the vectors $\mathbf{X}_j$, if and only if a vector $\mathbf{g}$ exists such that

$$\mathbf{V} = X\mathbf{g}$$

The minimum of the quantity $|\mathbf{Y} - \mathbf{V}|^2$ can be obtained if and only if $\mathbf{V}$ is the orthogonal projection of $\mathbf{Y}$ onto U_s. Let this minimizing vector be denoted by $\hat{\mu}$. Since $\hat{\mu}$ is an element of U_s, we may write

$$\hat{\mathbf{\mu}} = g_0 \mathbf{X}_0 + \cdots + g_k \mathbf{X}_k$$

for some set of coefficients $\{g_0, \ldots, g_k\}$. Now $\hat{\mu}$ is a function of $\mathbf{Y}$ alone, and therefore the coefficients $\{g_0, \ldots, g_k\}$ may also be taken to be functions of $\mathbf{Y}$ alone. Hence they are least-squares estimators whose existence has been proved.

Any $\{g_0, \ldots, g_k\}$ which depend on $\mathbf{Y}$ alone will be least-squares estimators if and only if $X\mathbf{g} = \hat{\mu}$, that is, if and only if

$$\begin{aligned}
\mathbf{Y} - X\mathbf{g} &\perp U_s \\
\mathbf{Y} - X\mathbf{g} &\perp X_j & j &= 0, \ldots, k \\
\mathbf{X}'_j(\mathbf{Y} - X'\mathbf{g}) &= 0 & j &= 0, \ldots, k \\
X(\mathbf{Y} - X'\mathbf{g}) &= 0 \\
X'X\mathbf{g} &= X'\mathbf{Y}
\end{aligned} \tag{6}$$

which are the normal equations. Hence the lemma is proved.

EXAMPLE 1

Assume a regression model with $k = 1$ and $m = 2$. Find the least-squares estimate of γ.

Solution:

$$\hat{\gamma} = (X'X)^{-1}X'\mathbf{Y}$$

where

$$X = \begin{bmatrix} 1 & x_1 \\ 1 & x_2 \end{bmatrix} \quad \text{and} \quad \mathbf{Y} = \begin{bmatrix} y_1 \\ y_2 \end{bmatrix}$$

Therefore

$$X'X = \begin{bmatrix} 2 & \Sigma x_i \\ \Sigma x_i & \Sigma x_i^2 \end{bmatrix}$$

$$X'\mathbf{Y} = \begin{bmatrix} \Sigma y_i \\ \Sigma x_i y_i \end{bmatrix}$$

and

$$(X'X)^{-1} = \frac{1}{2\Sigma x_i^2 - (\Sigma x_i)^2} \begin{bmatrix} \Sigma x_i^2 & -\Sigma x_i \\ -\Sigma x_i & 2 \end{bmatrix}$$

So

$$\hat{\gamma} = \frac{1}{2\Sigma x_i^2 - (\Sigma x_i)^2} \begin{bmatrix} \Sigma x_i^2 & -\Sigma x_i \\ -\Sigma x_i & 2 \end{bmatrix} \begin{bmatrix} \Sigma y_i \\ \Sigma x_i y_i \end{bmatrix}$$

or

$$\hat{\gamma}_0 = \frac{\Sigma y_i \Sigma x_i^2 - \Sigma x_i \Sigma x_i y_i}{2\Sigma x_i^2 - (\Sigma x_i)^2}$$

$$\hat{\gamma}_1 = \frac{-\Sigma x_i \Sigma y_i + 2\Sigma x_i y_i}{2\Sigma x_i^2 - (\Sigma x_i)^2}$$

Definition 1

Any linear function $\beta = \mathbf{b}'\boldsymbol{\gamma}$ of the unknown parameters $\{\gamma_0, \ldots, \gamma_k\}$ with known constant coefficients $\{b_0, \ldots, b_k\}$ will be called a *parametric function*. If there exists an unbiased linear estimate of β, then β will be called an *estimable parametric function*.

Lemma 2

A parametric function $\beta = \mathbf{b}'\boldsymbol{\gamma}$ is estimable if and only if $\mathbf{b}' = \boldsymbol{\alpha}'X$, that is, if and only if $\mathbf{b}'$ is a linear combination of the columns of X.

Proof:

From the above definition, β is estimable if and only if there exists a vector $\boldsymbol{\alpha}$ such that $E(\boldsymbol{\alpha}'\mathbf{Y}) = \beta$. However, $E(\boldsymbol{\alpha}'\mathbf{Y}) = \boldsymbol{\alpha}'X\boldsymbol{\gamma} = \mathbf{b}'\boldsymbol{\gamma}$. Thus β is estimable if and only if $\boldsymbol{\alpha}'X = \mathbf{b}'$.

Lemma 3

Let U_s be the space spanned by the columns of X. Then, if β is estimable, there exists a unique unbiased linear estimator of β, say $\boldsymbol{\alpha}^{*\prime}\mathbf{Y}$, where $\boldsymbol{\alpha}^*$ is in U_s. Furthermore, $\boldsymbol{\alpha}^*$ is the projection of $\boldsymbol{\alpha}$ on U_s, where $\boldsymbol{\alpha}'\mathbf{Y}$ is any other unbiased estimate of β.

Proof:

Since β is an estimable function, there exists an $\boldsymbol{\alpha}$ such that $E(\boldsymbol{\alpha}'\mathbf{Y}) = \beta$. If $\boldsymbol{\alpha} = \boldsymbol{\alpha}^* + \mathbf{g}$, where $\mathbf{g} \perp U_s$ and $\boldsymbol{\alpha}^* \in U_s$, then

$$\beta = E(\boldsymbol{\alpha}'\mathbf{Y}) = E(\boldsymbol{\alpha}^{*\prime}\mathbf{Y}) + E(\mathbf{g}' \times \boldsymbol{\gamma}) = E(\boldsymbol{\alpha}^{*\prime}\mathbf{Y})$$

since $\mathbf{g}'X = 0$. Thus $\boldsymbol{\alpha}^{*\prime}\mathbf{Y}$ is a linear unbiased estimate of β. If $\boldsymbol{\alpha}_0'\mathbf{Y}$ is another such estimate, then

$$0 = E(\boldsymbol{\alpha}^{*\prime}\mathbf{Y}) - E(\boldsymbol{\alpha}_0'\mathbf{Y}) = (\boldsymbol{\alpha}^* - \boldsymbol{\alpha}_0)'X\boldsymbol{\gamma}$$

and $(\boldsymbol{\alpha}^* - \boldsymbol{\alpha}_0)'X = 0$. This implies that $\boldsymbol{\alpha}^* - \boldsymbol{\alpha}_0$ is both in U_s and perpendicular to U_s. Hence $\boldsymbol{\alpha}_0 = \boldsymbol{\alpha}^*$, proving the uniqueness of $\boldsymbol{\alpha}^{*\prime}\mathbf{Y}$. The fact that $\boldsymbol{\alpha}^*$ is the projection of $\boldsymbol{\alpha}$ on U_s for any unbiased estimator $\boldsymbol{\alpha}'\mathbf{Y}$ can be inferred from the earlier part of the proof.

Let us now consider the Gauss-Markoff theorem for least-squares estimators.

Theorem 1

Let $E(\mathbf{Y}) = X\boldsymbol{\gamma}$ and $\Sigma_Y = \sigma^2 I$. Then there exists a unique unbiased linear estimator $\hat{\beta}$ for every estimable function $\beta = \mathbf{b}'\boldsymbol{\gamma}$. Furthermore, $\hat{\beta}$ has minimum variance in the class of all unbiased linear estimates. It may be obtained from β by replacing the $\{\gamma_0, \ldots, \gamma_k\}$ by any set of least-squares estimators $\{\hat{\gamma}_0, \ldots, \hat{\gamma}_k\}$.

Proof:

Lemma 3 gives the existence and uniqueness of the linear unbiased estimator $\boldsymbol{\alpha}^{*\prime}\mathbf{Y}$ of β, where $\boldsymbol{\alpha}^* \in U_s$. If $\boldsymbol{\alpha}'\mathbf{Y}$ is any other unbiased estimator, then $\boldsymbol{\alpha}^*$ is the projection of $\boldsymbol{\alpha}$ on U_s. That is,

$$|\boldsymbol{\alpha}|^2 = |\boldsymbol{\alpha}^*|^2 + |\boldsymbol{\alpha} - \boldsymbol{\alpha}^*|^2 \tag{7}$$

Thus

$$\begin{aligned} \operatorname{var}(\boldsymbol{\alpha}'\mathbf{Y}) &= \boldsymbol{\alpha}'\Sigma_Y\boldsymbol{\alpha} = \sigma^2|\boldsymbol{\alpha}|^2 \\ &= \sigma^2|\boldsymbol{\alpha}^*|^2 + \sigma^2|\boldsymbol{\alpha} - \boldsymbol{\alpha}^*|^2 \\ &= \operatorname{var}(\boldsymbol{\alpha}^{*\prime}\mathbf{Y}) + \sigma^2|\boldsymbol{\alpha} - \boldsymbol{\alpha}^*|^2 \end{aligned} \tag{8}$$

Equation (8) shows that var $(\alpha'Y) \geq$ var $(\alpha^{*'}Y)$ with equality holding only if $\alpha = \alpha^*$. This proves that $\alpha^{*'}Y$ is the unique linear unbiased estimator of β with minimum variance.

We must now show that $\alpha^{*'}Y = b'\hat{\gamma}$. If $\hat{\mu} = X\hat{\gamma}$ is the projection of Y on U_s, then $\alpha^{*'}(Y - \hat{\mu}) = 0$ since Y is in U_s and $Y - \hat{\mu}$ is perpendicular to U_s. Also $b' = \alpha^{*'}X$ since β is estimable. Therefore $\alpha^{*'}Y = \alpha^{*'}\hat{\mu} = \alpha^{*'}X\hat{\gamma} = b'\hat{\gamma}$, and the theorem is proved.

EXAMPLE 2

If $\beta_1, \ldots, \beta_r$ are estimable functions and $\hat{\beta}_i = \sum_{j=0}^{k} b_{ij}\hat{\gamma}_j$ is the least-squares estimate of β_i, prove that every linear combination $\beta = \sum_{i=1}^{r} n_i \beta_i$ is estimable and that its least-squares estimate is $\sum_{i=1}^{r} n_i \hat{\beta}_i$.

Solution:

Let $\beta_i = \sum_{j=0}^{k} b_{ij}\gamma_j$. Suppose that β is estimable. Then

$$\beta = \sum_{i=1}^{r} n_i \left(\sum_{j=0}^{k} b_{ij}\gamma_j \right) = \sum_{j=0}^{k} \left(\sum_{i=1}^{r} n_i b_{ij} \right) \gamma_j$$

Thus

$$\beta = \sum_{j=0}^{k} b'_j \gamma_j \qquad \text{where } b'_j = \sum_{i=1}^{r} n_i b_{ij}$$
$$\text{for } j = 0, \ldots, k$$

Applying the above theorem to this expression for β, we find that

$$\hat{\beta} = \sum_{j=0}^{k} b'_j \hat{\gamma}_j$$

where the $\{\hat{\gamma}_j\}$ are any set of least-squares estimates of the $\{\gamma_j\}$; so that

$$\hat{\beta} = \sum_{j=0}^{k} \left(\sum_{i=1}^{r} n_i b_{ij} \right) \hat{\gamma}_j$$

Applying the previous theorem to each β_i gives

$$\hat{\beta}_i = \sum_{j=0}^{k} b_{ij}\hat{\gamma}_j$$

These last two equations produce the relationship

$$\hat{\beta} = \sum_{i=1}^{r} n_i \hat{\beta}_i$$

It remains to be proved that β is indeed estimable. Express β as $\mathbf{n'B}$, where $\mathbf{n'} = (n_1, \ldots, n_r)$ and $\mathbf{B'} = (\beta_1, \ldots, \beta_r)$. But each $\beta_i = E(\boldsymbol{\alpha}_i'\mathbf{Y})$, so that

$$\beta = (n_i, \ldots, n_r) \begin{bmatrix} E(\boldsymbol{\alpha}_1'\mathbf{Y}) \\ \cdot \\ \cdot \\ \cdot \\ E(\boldsymbol{\alpha}_r'\mathbf{Y}) \end{bmatrix}$$
$$= E(n_1\boldsymbol{\alpha}_1'\mathbf{Y}) + \cdots + E(n_r\boldsymbol{\alpha}_r'\mathbf{Y})$$
$$= (n_1\boldsymbol{\alpha}_1' + \cdots + n_r\boldsymbol{\alpha}_r')E(\mathbf{Y})$$

Hence there exists an unbiased linear estimator for β, and so β is estimable.

Correlated case

Let $Y_1, \ldots, Y_n$ be correlated; i.e., let the covariance matrix $\boldsymbol{\Sigma}_Y$ be an arbitrary, symmetric, positive definite matrix.

Definition 2

An unbiased linear estimator $m(y)$ with covariance matrix $\boldsymbol{\Sigma}_m$ is said to be minimum variance if $\boldsymbol{\Sigma}_m \leq \boldsymbol{\Sigma}_q$ for all linear unbiased estimators $q(y)$. We say that $\boldsymbol{\Sigma}_m \leq \boldsymbol{\Sigma}_q$ if $\boldsymbol{\Sigma}_q - \boldsymbol{\Sigma}_m$ is nonnegative definite.

Theorem 2 Gauss-Markoff Theorem

If $E(\mathbf{Y}) = X\boldsymbol{\gamma}$, $\boldsymbol{\Sigma}_Y$ is nonsingular, and the rank of X is $k+1$, then the Markoff estimators

$$\hat{\boldsymbol{\gamma}} = (X'\boldsymbol{\Sigma}_Y^{-1}X)^{-1}X'\boldsymbol{\Sigma}_Y^{-1}\mathbf{Y} \tag{9}$$

are minimum-variance-unbiased linear estimators of the regression coefficients of Eq. (1).

Proof:

To show that the estimators are unbiased, we have

$$\begin{aligned} E(\hat{\gamma}) &= E[(X'\Sigma_Y^{-1}X)^{-1}X'\Sigma_Y^{-1}\mathbf{Y}] \\ &= (X'\Sigma_Y^{-1}X)^{-1}X'\Sigma_Y^{-1}X\gamma \\ &= \gamma \end{aligned} \quad (10)$$

since $E(\mathbf{Y}) = X\gamma$ by hypothesis.

To show that the Markoff estimates are unbiased, assume that RY is an unbiased linear estimator of γ. Then $\Sigma_R = R\Sigma_Y R'$, and using the fact that $RX = I$, we may write Σ_R in the following identity:[1]

$$\begin{aligned} R\Sigma_Y R' &= [(X'\Sigma_Y^{-1}X)^{-1}X'\Sigma_Y^{-\frac{1}{2}}][(X'\Sigma_Y^{-1}X)^{-1}X'\Sigma_Y^{-\frac{1}{2}}]' \\ &\quad + [R\Sigma_Y^{\frac{1}{2}} - (X'\Sigma_Y^{-1}X)^{-1}X'\Sigma_Y^{-\frac{1}{2}}][R\Sigma_Y^{\frac{1}{2}} - (X'\Sigma_Y^{-1}X)^{-1}X'\Sigma_Y^{-\frac{1}{2}}]' \end{aligned} \quad (11)$$

where we denote by $\Sigma^{\frac{1}{2}}$ the unique positive definite square root of the matrix Σ. Each of the terms on the right of the identity (11) is a non-negative definite matrix. A strict minimum for $R\Sigma_Y R'$ occurs only when the second term is zero, i.e., when

$$R = (X'\Sigma_Y^{-1}X)^{-1}X'\Sigma_Y^{-1}$$

Thus the Markoff estimators have minimum variance.

Corollary

The Markoff estimators minimize the generalized variance in the class of unbiased linear estimators.

Proof:

The generalized variance of an estimator is the determinant of its covariance matrix. From the theorem we have that $\Sigma_q \geq \Sigma_m$, and we must show that $|\Sigma_q| \geq |\Sigma_m|$. Since Σ_m is nonsingular and positive semidefinite, it is positive definite. Then Σ_q and Σ_m can be simultaneously diagonalized by a nonsingular transformation. If D_m and D_q are the diagonalized matrices, then $D_q \geq D_m$. It follows that $D_q - D_m$ is nonnegative definite and each diagonal element of D_q is greater than the corresponding element of D_m. Hence $|D_q| \geq |D_m|$ or $|\Sigma_q| \geq |\Sigma_m|$, and the Markoff estimators minimize the generalized variance in the class of unbiased estimators.

EXAMPLE 3

Let $y_t = x_t + \mu_t$ be a stochastic process observed at $t = 1, \ldots, n$ with mean value $\mu_t = E(y_t)$. If it is assumed that each μ_t is given by a linear expression in some unknown parameters $\theta_1, \ldots, \theta_k$, where $k < n$, that is, if

$$\mathbf{\mu} = T\mathbf{\theta} \quad \text{where } \mathbf{\mu}' = (\mu_1, \ldots, \mu_n)$$

$\mathbf{\theta}' = (\theta_1, \ldots, \theta_k)$, and T is an $n \times k$ matrix of known constants and of rank k, then using the method of least squares to find estimates of the θ_i leads us to observe that the Markoff estimates $\hat{\gamma}_i$ provide a minimum-variance-unbiased estimate of the θ_i, where

$$\hat{\gamma} = \begin{bmatrix} \hat{\theta}_1 \\ \cdot \\ \cdot \\ \cdot \\ \hat{\theta}_n \end{bmatrix} = (T'\Sigma_Y^{-1}T)^{-1}T'\Sigma_Y^{-1}\mathbf{Y}$$

3 □ THE ESTIMATION OF LOCATION AND SCALE PARAMETERS OF A DISTRIBUTION

The generalized least-squares principle is introduced for estimating the location and scale parameters of a distribution of the form $F[(x - \mu)/\sigma]$ based on order statistics. The results are due to Lloyd [3] and Downtown [1].

Suppose we have n independent observations $X_1, X_2, \ldots, X_n$ on the continuous random variable X whose distribution is $F[(x - \mu)/\sigma]$, where μ and σ are the location and scale parameters, respectively, and let $x_{(1)} < x_{(2)} < \cdots < x_{(n)}$ be n-order statistics, arranged in order of magnitude. Let

$$z_{(i)} = \frac{x_{(i)} - \mu}{\sigma} \quad i = 1, 2, \ldots, n$$

and

$$\begin{aligned} E(z_{(i)}) &= \alpha_i \\ V(z_{(i)}) &= v_{ii}\sigma^2 \\ \text{cov}(z_{(i)}, z_{(j)}) &= v_{ij}\sigma^2 \end{aligned} \tag{1}$$

so that

$$A = \begin{bmatrix} 1 & \alpha_1 \\ 1 & \alpha_2 \\ \cdots & \cdots \\ 1 & \alpha_n \end{bmatrix} \quad \theta = \begin{bmatrix} \mu \\ \sigma \end{bmatrix} \tag{2}$$

and

$$V = (v_{jj}) \quad \boldsymbol{\alpha} = \begin{bmatrix} \alpha_1 \\ \alpha_2 \\ \cdot \\ \cdot \\ \cdot \\ \alpha_n \end{bmatrix} \quad 1 = \begin{bmatrix} 1 \\ 1 \\ \cdot \\ \cdot \\ \cdot \\ 1 \end{bmatrix}$$

From (1) it is clear that

$$\begin{aligned} E(\mathbf{X}) &= \mu 1 + \sigma \boldsymbol{\alpha} \\ \text{var}(\mathbf{X}) &= \sigma^2 V \end{aligned} \tag{3}$$

For the least-squares estimate of μ and σ, we minimize the quadratic form

$$(\mathbf{X} - \mathbf{A}\boldsymbol{\theta})' V^{-1} (\mathbf{X} - \mathbf{A}\boldsymbol{\theta}) \tag{4}$$

with respect to $\boldsymbol{\theta}$, which yields

$$\begin{aligned} \mu^* &= -\boldsymbol{\alpha}' \Gamma \mathbf{X} \\ \sigma^* &= 1' \Gamma \mathbf{X} \end{aligned} \tag{5}$$

where

$$\begin{aligned} \Gamma &= \Omega(1\alpha' - \alpha'1)\frac{\Omega}{\Delta} \\ \Delta &= (1'\Omega 1)\alpha(\alpha'\Omega\alpha) - \lambda(1'\Omega\alpha)^2 \\ \Omega &= v^{-1} \end{aligned}$$

The variances and the covariance of the estimates are given by

$$\begin{aligned} \text{var } \mu^* &= \frac{(\alpha'\Omega\alpha)\sigma^2}{\Delta} \\ \text{var } \sigma^* &= \frac{(1'\Omega 1)\sigma^2}{\Delta} \\ \text{cov } (\mu^*, \sigma^*) &= -\frac{(1'\Omega\alpha)\sigma^2}{\Delta} \end{aligned} \tag{6}$$

The estimates possess the minimum-variance property among the class of linear estimates of μ and σ.

If the parent distribution is symmetrical, the calculation is simplified since

$$E(Z) = \alpha = \begin{bmatrix} \alpha_1 \\ \cdot \\ \cdot \\ \cdot \\ \alpha_n \end{bmatrix}$$

has $\alpha_i = -\alpha_{n+1-i}$ for all i, so that

$$\alpha'\Omega 1 = 1'\Omega\alpha = 0$$

$$\mu^* = \frac{1'\Omega X}{1'\Omega 1} \tag{7}$$

$$\sigma^* = \frac{\alpha'\Omega X}{\alpha'\Omega\alpha}$$

and the variances simplify to

$$\text{var } \mu^* = \frac{\sigma^2}{1'\Omega 1}$$

$$\text{var } \sigma^* = \frac{\sigma^2}{\alpha'\Omega\alpha} \tag{8}$$

$$\text{cov } (\mu^*, \sigma^*) = 0$$

Sometimes it is desired to obtain estimates of μ and σ based on few selected order statistics. The above techniques are then useful, and all calculation is based on the selected order statistics. For example, suppose $X_{(n_1)} < X_{(n_2)} < \cdots < X_{(n_k)}$ be k selected order statistics with ranks $n_1, n_2, \ldots, n_k$ satisfying the relations $1 \leq n_1 < n_2 < \cdots < n_k \leq n$ so that

$$E(X_{(n_i)}) = \mu + \sigma\alpha_{(n_i)}$$
$$\text{var } X_{(n_i)} = \sigma^2 \, v_{n_i n_i} \tag{9}$$
$$\text{cov } (X_{(n_i)}, X_{(n_j)}) = \sigma^2 \, v_{n_i, n_j}$$

Then, defining

$$\alpha = \begin{bmatrix} \alpha_{n_1} \\ \alpha_{n_2} \\ \cdot \\ \cdot \\ \cdot \\ \alpha_{n_k} \end{bmatrix}$$

$$V = (v_{n_i, n_j})_{k \times k}$$

we obtain similar expressions to those in (5) and (6) for the estimates and the variances and covariance of the estimates, except for the dimensions and elements of **α** and V. In such an estimation procedure the related problem of choosing a suitable set of k optimum order statistics arises; it is solved in certain cases by minimizing the generalized variance of the estimates which are given by

$$\Lambda = \text{var } \mu^* \text{ var } \sigma^* - \text{cov}^2(\mu^*, \sigma^*) \tag{10}$$

with respect to the choice of $n_1, n_2, \ldots, n_k$. For such problems the readers are referred to Ogawa [7], Kulldorff [2], Saleh and Ali [5], and Sarhan and Greenberg [4].

The main difficulty in the application of the above theory arises in the computation of the exact variance-covariance matrix of order statistics. Therefore, one takes recourse to large-sample theory.

4 □ AN ILLUSTRATION

Consider the two-parameter exponential distribution given by

$$dF(x) = \frac{1}{\sigma} e^{-(x-u)/\sigma} dx \qquad x \geq \mu, \, \sigma > 0 \tag{11}$$

and assume that a large sample of size n has been drawn and that k ($\leq n$) is given. Let the ordered observations in a random sample of size n be $x_{(1)} < x_{(2)} < \cdots < x_{(n)}$ and consider the k sample quantiles $x_{(n_1)}, \ldots, x_{(n_k)}$, where $n_1, n_2, \ldots, n_k$ are the ranks which satisfy the relation $1 \leq n_1 < n_2 < \cdots < n_k \leq n$ and are determined by k fixed real numbers $0 < p_1 < p_2 < \cdots < p_k < 1$ and $n_i = [np_i] + 1, i = 1, 2, \ldots, k$. In the Euler notation $[np_i]$ denotes the largest integer contained in the brackets. Let $p_0 = 0$ and $p_{k+1} = 1$. The asymptotic distribution of $X_{(n_1)}, X_{(n_2)}, \ldots, X_{(n_k)}$ is a k-variate normal distribution (Ogawa [7]) with means $(\mu + \sigma u_1, \mu + \sigma u_2, \ldots, \mu + \sigma u_k)$ and dispersion matrix $V = (v_{ij})\sigma^2/n$, with $v_{ii} = v_{ij} = v_{ji} = (e^{u_i} - 1)$, $i < j$, $i, j = 1, 2, \ldots, k$. The elements of the inverse $V^{-1} = \Omega$ are

$$v'_{ii} = \frac{(e^{-u_i-1} - e^{-u_i+1}) e^{-2\mu_i}}{(e^{-\mu_i} - e^{-\mu_{i+1}})(e^{-u_{i+1}} - e^{-u_i})}$$

$$v'_{i,i-1} = v'_{i-1,i} = \frac{1}{e^{-\mu_i} - e^{-u_{i-1}}} \tag{12}$$

$$v_{ij} = 0 \qquad \text{for } |i - j| > 1$$

where $u_i = \ln(1 - p_i)^{-1}$, $i = 1, 2, \ldots, k$, are the quantiles of the standardized exponential distribution e^{-u} corresponding to $p_1, p_2, \ldots, p_k$. Now by the application of the generalized least-squares principle, the asymptotically best linear unbiased estimates of μ and σ based on the k quantiles are given by

$$\mu^* = x_{(n_1)} - \sigma^* \mu_1$$
$$\sigma^* = \sum_{i=1}^{k} b_i x_{(n_i)}$$

when

$$b_1 = -\frac{u_2 - u_1}{(e^{u_2} - e^{u_1})L}$$
$$b_i = \frac{1}{L}\left(\frac{u_i - u_{i-1}}{e^{u_i} - e^{u_{i-1}}} - \frac{u_{i+1} - u_i}{e^{u_{i-1}} - e^{u_i}}\right) \quad i = 2, 3, \ldots, k-1$$
$$b_k = \frac{1}{L}\frac{u_k - r_{k-1}}{e^{u_n} - e^{u_{k-1}}} \quad (13)$$
$$L = \sum_{i=2}^{k} \frac{(u_i - u_{i-1})^2}{e^{u_i} - e^{u_{i-1}}}$$

The variances and covariances of the estimates are

$$\operatorname{var} \mu^* = \left[\left(\frac{u_1^2}{L}\right) + (e^{u_1} - 1)\right]\frac{\sigma^2}{n}$$
$$\operatorname{var} \sigma^* = \frac{1}{L}\frac{\sigma^2}{n} \quad (14)$$
$$\operatorname{cov}(\mu^*, \sigma^*) = \frac{u_1}{L}\frac{\sigma^2}{n}$$

The generalized variance of the estimate is

$$\Lambda = \operatorname{var} \mu^* \operatorname{var} \sigma^* - \operatorname{cov}^2(\mu^*, \sigma^*) = \frac{e^{u_1} - 1}{L}\frac{\sigma^4}{n^2} \quad (15)$$

If $\mu = 0$, then the estimate of σ is given by

$$\sigma^* = \sum_{i=1}^{k} b_i X_{(n_i)}$$

with variance

$$\operatorname{var} \sigma^* = Q_k^{-1} \frac{\sigma^2}{n} \tag{16}$$

where

$$Q_k = \sum_{i=1}^{k} \frac{(u_i - u_{i-1})^2}{e^{u_i} - e^{u_{i-1}}}$$

The subscript of k is the dimension of the function Q defined above. It may be noted that the above estimates possess the minimal-variance property among the class of all linear estimates based on the given fixed quantiles $X_{(n_1)}, \ldots, X_{(n_k)}$.

5 □ DETERMINATION OF OPTIMUM QUANTILES FOR THE BEST LINEAR UNBIASED ESTIMATES OF σ WHEN $\mu = 0$

In order to determine the k optimum quantiles, we have to choose among $\binom{n}{k}$ possible sets, a particular set such that the linear estimates based on them possess minimum possible variance. This is possible if we minimize the variance $Q_k^{-1}\sigma^2/n$, or, equivalently, if we maximize Q_k with respect to $u_1, u_2, \ldots, u_k$. Suppose the maximization of Q_k with respect to $u_1, u_2, \ldots, u_k$ occurs at $(u_1^0, u_2^0, \ldots, u_k^0)$. Then, $u_1^0, \ldots, u_k^0$ will determine the optimum spacings $p_1^0, p_2^0, \ldots, p_k^0$ of the quantiles (order statistics) by the relations

$$p_i^0 = 1 - e^{-u_i^0} \qquad i = 1, 2, \ldots, k$$

for the best linear unbiased estimate of σ of the exponential distribution based on k optimum quantiles $x_{(n_1^0)}, \ldots, x_{(n_k^0)}$, where the ranks are given by

$$n_i^0 = [np_i^0] + 1 \qquad i = 1, 2, \ldots, k$$

The asymptotic relative efficiency (ARE) of the estimate of σ with respect to the maximum-likelihood estimate (MLE) using all the observations is

$$\operatorname{ARE} \sigma^* = Q_k^* \sum_{i=1}^{k} \frac{(u_i - u_{i-1})^2}{e^{u_i} - e^{u_{i-1}}}$$

32 PARAMETRIC ESTIMATION

Let S_{MLE} be the MLE of σ and S_L be a linear estimate of σ based on n observations; then

$$\text{ARE} = \lim_{n \to \infty} \frac{\text{var } S_{MLE}}{\text{var } S_L}$$

where $Q_k{}^*$ has been computed based on $u_1{}^0, u_2{}^0, \ldots, u_k{}^0$.

The least-squares principle along with the asymptotic theory using the quantiles is applicable to a large class of distributions of which the relevant ones are given in the accompanying table.

Distribution	Density Function	
Normal	$\dfrac{1}{\sqrt{2\pi}} e^{-x^2/2}$	$-\infty < x < \infty$
Exponential	e^{-x}	$(x > 0)$
Gamma	$\dfrac{x^{p-1}}{\Gamma p} e^{-x}$	$x > 0, p > 0$
Log normal	$\dfrac{1}{\sigma x \sqrt{2\pi}} e^{-(\log x)^2/2\sigma^2}$	$x > 0, \sigma > 0$
Logistic	$e^{-x}(1 + e^{-x})^{-2}$	
Weibull	$\dfrac{1}{s} x^{1/s - 1} e^{-x^{1/s}}$	$(x > 0), s > 0$
Extreme value	$e^{-x - e^{-x}}$	

The shape parameters p, σ, and s of the gamma, log normal, and Weibull distributions, respectively, are assumed known.

PROBLEMS

1. (Lloyd) For the estimation of μ and σ of the rectangular distribution

$$dF(x) = \frac{dx}{\sigma} \qquad \mu - \tfrac{1}{2}\sigma \leq x \leq \mu + \tfrac{1}{2}\sigma$$

$$\alpha_{(i)} = E\left[\frac{x_{(i)}}{\sigma}\right] = \frac{i}{n+1} - \frac{1}{2}$$

and the elements of the dispersion matrix V are

$$v_{ij} = \frac{i(n-j+1)}{(n+1)^2(n-2)} \qquad i \leq j$$

the inverse of V is

$$\Omega = \begin{bmatrix} 2 & -1 & 0 & 0 & 0 & \cdots & 0 & 0 \\ -1 & 2 & -1 & 0 & 0 & \cdots & 0 & 0 \\ 0 & -1 & 2 & -1 & 0 & \cdots & 0 & 0 \\ \cdot & \cdot & \cdot & \cdot & \cdot & \cdots & \cdot & \cdot \\ \cdot & \cdot & \cdot & \cdot & \cdot & \cdots & \cdot & \cdot \\ \cdot & \cdot & \cdot & \cdot & \cdot & \cdots & \cdot & \cdot \\ 0 & 0 & 0 & 0 & 0 & \cdots & -1 & 2 \end{bmatrix}$$

and the estimates are

$$\mu^* = \tfrac{1}{2}(x_{(1)} + x_{(n)})$$

$$\sigma^* = \frac{(n+1)(x_{(n)} - x_{(1)})}{n-1}$$

$$\operatorname{var} \mu^* = \frac{\sigma^2}{2(n+1)(n+2)}$$

$$\operatorname{var} \sigma^* = \frac{2\sigma^2}{(n-1)(n+2)}$$

$$\operatorname{cov}(\mu^*, \sigma^*) = 0$$

2. (Lloyd) Show that the least-squares estimate of μ satisfies $\operatorname{var}(\mu^*) \leq \sigma^2/n$ with respect to its variance.

3. (Kulldorff) For the exponential distribution

$$dF(x) = \frac{1}{\sigma} e^{-(x-\mu)/\sigma} \, dx \qquad x \geq \mu$$

$$\sigma > 0$$

Show that

$$E(X_{(n_i)}) = \sum_{j=n_i-1}^{n_{i+1}-1} (n-j+1)^{-1}$$

$$\operatorname{cov}(X_{(n_i)}, X_{(n_j)}) = \operatorname{var} X_{(n_i)} = \sum_{j=n_i-1}^{n_{i+1}-1} (n-j+1)^{-2} \qquad n_i < n_j$$

where n_i is the rank of the order statistic $X_{(n_i)}$.

4. (Sarhan) For the above problem, using all the order statistics, show that the variance-covariance matrix has the inverse given by

$$V^{-1} = \begin{bmatrix} n^2 + (n-1)^2 & -(n-1)^2 & 0 & \cdots & 0 & 0 & 0 \\ -(n-1)^2 & (n-1)^2 + (n-2)^2 & -(n-2)^2 & \cdots & 0 & 0 & 0 \\ \cdot & \cdot & \cdot & \cdots & \cdot & \cdot & \cdot \\ \cdot & \cdot & \cdot & \cdots & \cdot & \cdot & \cdot \\ 0 & 0 & 0 & \cdots & -2 & 2^2 + 1^2 & -1^2 \\ 0 & 0 & 0 & \cdots & 0 & -1^2 & 1^2 \end{bmatrix}$$

and

$$\mu^* = X_{(1)} - \frac{\bar{X} - X_{(1)}}{n - 1}$$

$$\sigma^* = \frac{n(\bar{X} - X_{(1)})}{n - 1}$$

Compare with the maximum-likelihood estimators.

5. (*a*) (Saleh and Ali) For the exponential distribution show that optimum spacings for the k optimum quantiles for the estimate of μ and σ are

$$p_1^0 = \frac{1}{n + \frac{1}{2}}$$

$$p_{i+1}^0 = \frac{2 + (2n-1)\lambda_i^0}{2n+1} \quad i = 1, \ldots, k-1$$

where λ_i^0 $i = 1, 2, \ldots, k-1$, are optimum spacings for $k-1$ optimum quantiles for the estimation of σ alone when $\mu = 0$.

(*b*) Show that the asymptotic relative efficiency of the estimates of μ and σ are

$$\text{JAE }(\mu^*, \sigma^*) = \frac{(2n-1)^2}{2(n^2 - 1)} Q_{k-1}^0$$

$$\text{ARE }\sigma^* = \frac{2n-1}{2n+1} Q_{k-1}^0$$

$$\text{ARE }\mu^* = \frac{(2n-1)Q_{k-1}^0}{n\{(2n-1) \ln [(2n+1)/(2n-1)] + 2Q_{k-1}^0\}}$$

where Q_{k-1}^0 is the ARE of the estimate of σ based on $k - 1$ optimum quantiles and JAE stands for joint asymptotic efficiency, which can be defined in an obvious manner as ARE is defined.

(c) Prove that the optimum quantiles obtained above are unique.

REFERENCES

1   DOWNTOWN, F.: A Note on Ordered Least Squares Estimation, *Biometrika*, vol. 40, pp. 457–458, 1953.

2   KULLDORFF, G.: Estimation of One or Two Parameters of the Exponential Distribution on the Basis of Suitably Chosen Order Statistics, *Ann. Math. Statist.*, vol. 34, pp. 1419–1431, 1963.

3   LLOYD, E. H.: Least Squares Estimation of Location and Scale Parameters Using Order Statistics, *Biometrika*, vol. 39, pp. 88–95, 1952.

4   SARHAN, A. E., and B. G. GREENBERG: "Contribution to Order Statistics," John Wiley & Sons, Inc., New York, 1962.

5   SALEH, A. K., MD. EHSANES, and MIR M. ALI: Asymptotic Optimum Quantiles for the Estimation of the Parameters of the Negative Exponential Distributions, *Ann. Math. Statist.*, vol. 37, pp. 143–151, 1966.

6   SCHEFFÉ, H.: "Analysis of Variance," John Wiley & Sons, Inc., New York, 1958.

7   OGAWA, J.: Contributions to the Theory of Systematic Statistics I, *Osaka Math. J.*, vol. 3, pp. 131–142, 1951.

chapter Three

Optimal Spacing of a Regressor Variable

1 □ INTRODUCTION

The author's experience with applied industrial problems has demonstrated that the choice of spacing of the given regressor variable in conducting experiments is very important. The proper choice not only avoids waste of material and manpower but also can result in reliable practical results.

The problem was considered by Smith [20] as early as 1918, and he could give a solution to the problem when the polynomial was of degree up to 6. One can also consider this as a problem of approximation theory (see Shohat and Tamarkin [19]).

We shall discuss the theorem of De la Garza [3], which is a consequence of the moment problem and illustrate it by an example. Then we append the minimax variance result given by Hoel [8] and Guest [5]. The recent results of Hoel and Levine [11] are stated as problems for solution.

A general problem of optimum design has been considered by Kiefer and Wolfowitz [14, 15].

2 □ DE LA GARZA THEOREM

The main result of this section is due to De la Garza [3]. It is that the information contained in n uncorrelated observations of the real-valued random variable Y is the same whether the observations are made at a distinct location x_i, $i = 1, \ldots, n$, $n > k + 1$, or at only $k + 1$ distinct locations r_j, $j = 1, \ldots, k + 1$. Before proving this we make several definitions and preliminary relations and we impose some restrictions on the problem.

Definitions

The information of Y_i is $w_i = 1/(\text{var } Y_i)$. The total information of the vector Y is $Q = \sum_i w_i = \text{trace } \Sigma_Y^{-1}$. In this section $\sum_i$ denotes a summation over $i = 1, \ldots, n$. For a particular set $\{x_i, i = 1, \ldots, n\}$, called a *spacing*, the information matrix is $X' \Sigma_Y^{-1} X$. The form of this matrix is displayed in Eq. (5). The total information of a spacing is thus the (1,1) element of the information matrix.

Preliminary relations

As before, vector quantities are boldface.

Let $P(x)$ be a polynomial of degree k, $k \geq 1$, with real coefficients,

$$P(x) = \alpha_0 + \alpha_1 x + \alpha_2 x^2 + \cdots + \alpha_k x^k$$

or

$$P(x) = \mathbf{x} \cdot \boldsymbol{\alpha}'$$

where

$$\mathbf{x} = (1, x, \ldots, x^k) \quad \text{and} \quad \boldsymbol{\alpha} = (\alpha_0, \alpha_1, \ldots, \alpha_k)$$

To represent $P(x)$ with the lagrangian interpolation formula, we choose

$k+1$ distinct real numbers $z_1, z_2, \ldots, z_{k+1}$. Then

$$P(x) = \frac{(x-z_2)(x-z_3) \cdots (x-z_{k+1})}{(z_1-z_2)(z_1-z_3) \cdots (z_1-z_{k+1})} P(z_1) + \cdots$$
$$+ \frac{(x-z_1)(x-z_2) \cdots (x-z_k)}{(z_{k+1}-z_1)(z_{k+1}-z_2) \cdots (z_{k+1}-z_k)} P(z_{k+1})$$

With obvious notation

$$P(x) = F(x,z_1)P(z_1) + F(x,z_2)P(z_2) + \cdots + F(x,z_{k+1})P(z_{k+1})$$

or

$$P(x) = \mathbf{F}(x,z) \cdot \mathbf{P}(z)' = \mathbf{x} \cdot \boldsymbol{\alpha} \qquad (1)$$

Let

$$Z = \begin{bmatrix} 1 & z_1 & \cdots & z_1^k \\ \cdots & \cdots & \cdots & \cdots \\ 1 & z_{k+1} & \cdots & z_{k+1}^k \end{bmatrix}$$

Since $P(z) = \mathbf{z} \cdot \boldsymbol{\alpha}'$, we obtain

$$P(z)' = Z\boldsymbol{\alpha}' \qquad (2)$$

Since Z is a Vandermonde form matrix with distinct elements, Z^{-1} exists and we can solve Eq. (2) for $\boldsymbol{\alpha}'$. That is,

$$\boldsymbol{\alpha}' = Z^{-1}\mathbf{P}(z)'$$

If we substitute this quantity into the right-hand side of Eq. (1), we obtain

$$\mathbf{x} \cdot \mathbf{z}^{-1}\mathbf{P}(z)' = \mathbf{P}(x,z) \cdot \mathbf{P}(z)'$$

This is an identity in $\mathbf{x}$, and we conclude that $\mathbf{x}Z^{-1} = \mathbf{F}(x,z)$. We can write this in matrix form

$$XZ^{-1} = \begin{bmatrix} F(x_1,z_1) & \cdots & F(x_1,z_{k+1}) \\ \cdots & \cdots & \cdots \\ F(x_n,z_1) & \cdots & F(x_n,z_{k+1}) \end{bmatrix} \qquad (3)$$

This is the required relation.

Restriction of the problem

Two restrictions are considered: first, max $x_i = 1$, and min $x_i = -1$; and second, the x_i, $i = 1, \ldots, n$, are distinct.

For the first, let the regressor variables be transformed by the linear relation

$$S_i = \frac{2x_i - x_n - x_1}{x_n - x_1} \tag{4}$$

where $x_1 < x_2 < \cdots < x_n$. The S_i are in the interval $[-1,1]$ and are proportional to the original regressor variables x_i. For simplicity we shall denote the S_i by x_i and consider henceforth that the x_i are in the closed interval $[-1,1]$ with min $x_i = -1$ and max $x_i = 1$.

For the second, suppose some of the x_i are not distinct, say

$$x_1 = x_2 = \cdots = x_p$$

with corresponding information $w_1, \ldots, w_p$. Then the information matrix for $w_1, w_2, \ldots, w_p, w_{p+1}, \ldots, w_n$ at $x_1, \ldots, x_p, x_{p+1}, \ldots, x_n$ is the same as the information matrix for $(w_1 + w_2 + \cdots + w_p)$, $w_{p+1}, \ldots, w_n$ at $x_p, x_{p+1}, \ldots, x_n$.

$$X'\Sigma_Y^{-1}X = \begin{bmatrix} \sum_i w_i & \sum_i w_i x_i & \cdots & \sum_i w_i x_i^k \\ \sum_i w_i x_i & \sum_i w_i x_i^2 & \cdots & \cdots \\ \vdots & \vdots & & \vdots \\ \sum_i w_i x_i^k & \sum_i w_i x_i^{k+1} & \cdots & \sum_i w_i x_i^{2k} \end{bmatrix}$$

$$= \begin{bmatrix} 1 & 1 & \cdots & 1 \\ x_p & x_{p+1} & \cdots & x_n \\ \vdots & \vdots & & \vdots \\ x_p^k & x_{p+1}^k & \cdots & x_n^k \end{bmatrix} \begin{bmatrix} \sum_{i=1}^p w_i & 0 & \cdots & 0 \\ 0 & w_{p+1} & \cdots & 0 \\ \vdots & & & \vdots \\ 0 & 0 & \cdots & w_n \end{bmatrix} \times$$

$$\begin{bmatrix} 1 & x_p & \cdots & x_p^k \\ 1 & x_{p+1} & \cdots & x_{p+1}^k \\ \vdots & \vdots & & \vdots \\ 1 & x_n & \cdots & x_n^k \end{bmatrix} \tag{5}$$

Thus if the x_i, $i = 1, \ldots, n$, are not distinct, we can group the observations and consider that all the x_i are distinct.

Theorem

Given a spacing of total information Q at n distinct locations x_i, $i = 1, \ldots, n$, $n > k + 1$, with min $x_i = -1$, max $x_i = 1$, it is always possible to respace Q at $k + 1$ distinct locations r_j, $j = 1, \ldots, k + 1$, in such

40 PARAMETRIC ESTIMATION

a way that $-1 \le r_j \le 1$ and

$$X'\Sigma_Y^{-1}X = R'UR \tag{6}$$

R is a $(k+1) \times (k+1)$ Vandermonde form matrix of the elements r_j, and $R'UR$ is the information matrix of the new spacing where n is divisible by $k+1$.

Proof:

We define a polynomial

$$P(r) = \alpha_0 + \alpha_1 r + \cdots + \alpha_k r^k + r^{k+1} \tag{7}$$

The coefficients are determined uniquely and nontrivially by the set of equations

$$(X'\Sigma_Y^{-1}X)\, \alpha' + \mathbf{f}' = 0 \tag{8}$$

where $\alpha = (\alpha_0, \ldots, \alpha_k)$, $\mathbf{f} = (f_{k+1}, \ldots, f_{2k+1})$, and $f_{k+L} = \sum_i w_i x_i^{k+L}$, $L = 1, 2, \ldots, k+1$.

We define a function u^*gh (which is proportional to the elements of the matrix U) by

$$u^*gh = \sum_i w_i \phi_i \prod_{\substack{j=1 \\ j \ne h \\ j \ne g}}^{k+1} (x_i - r_j) \qquad \begin{array}{l} g \ne h \\ g, h = 1, \ldots, k+1 \end{array} \tag{9}$$

where $\phi_i = \prod_{j=1}^{k+1} (x_i - r_j)$ and the r_j are the roots of the polynomial (7).

Now $u^*gh = 0$ if $g \ne h$ since Eq. (8) implies

$$\sum_i w_i \phi_1 x_i^L = 0 \qquad L = 0, 1, \ldots, k \tag{10}$$

and this obviously implies $u^*gh = 0$. Tos how that Eq. (8) implies Eq. (10) we write the $(L+1)$st equation from the set (8) as

$$\sum_i w_i x_i^L \alpha_0 + \sum_i w_i x_i^{L+1} \alpha_1 + \cdots + \sum_i w_i x_i^{L+k} \alpha_k + \sum_i w_i x_i^{L+k+1} = 0$$

Then

$$\sum_i w_i x_i^L (\alpha_0 + \alpha_1 x_i + \cdots + \alpha_k x_i^k + x_i^{k+1}) = 0$$

and

$$\sum_i w_i x_i^L \phi_i = 0$$

We now show that the roots of the polynomial (6) are real and distinct by showing that $u^*gh \neq 0$ if any of the roots are complex or repeated. Since the r_j are roots of a polynomial, any complex members of the set must appear as conjugate pairs. Suppose r_1 and r_2 are such a conjugate pair; let $r_1 = b_1 + b_2 i$ and $r_2 = b_1 - b_2 i$, $b_2 \neq 0$. We rewrite Eq. (9) in the form

$$u^*gh = \sum_i w_i (x_i - r_g)(x_i - r_h) \prod_{\substack{j=1 \\ j \neq g \\ j \neq h}}^{k+1} (x_i - r_j)^2 \qquad (11)$$

Hence

$$u_{12}^* = \sum_i w_i [(x_i - b_1)^2 + b_2^2] \prod_{j=3}^{k+1} (x_i - r_j)^2$$

All summands (there are at least $k + 2$ of them) in this summation are nonnegative; $u_{12}^* = 0$ requires that every summand be zero. However, $(x_i - b_1)^2 + b_2^2$ is never zero, and $\prod_{j=3}^{k+1} (x_i - r_j)^2$ is zero for at most $k - 1$ values of the x_i. Hence $u_{12}^* \neq 0$, a contradiction. We conclude that no roots may be complex since this procedure could be applied to all pairs of roots. To show that the roots are distinct, we suppose that $r_1 = r_2 = b_1$. Then Eq. (11) takes the form $u_{12}^* = \sum_i w_i (x_i - b_1)^2 \prod_{j=3}^{k+1} (x_i - r_j)^2$. Again each summand is nonnegative, and $u_{12}^* = 0$ requires that each summand be zero. There can be at most k of the x_i for which summands are zero; hence $u_{12}^* \neq 0$, a contradiction. As before, we conclude that all the roots are distinct.

We now show that the roots are in the closed interval $[-1,1]$. Suppose two roots, say r_1 and r_2, are outside the interval $[-1,1]$. Then the product, $(x_i - r_1)(x_i - r_2)$, is never zero and always has the same algebraic sign. Further,

$$u_{12}^* = \sum_i w_i (x_i - r_1)(x_i - r_2) \prod_{j=3}^{k+1} (x_i - r_j)^2 \qquad (12)$$

is the sum of more than $k + 1$ similarly signed summands. Thus $u_{12}^* = 0$ only if each summand is zero. Obviously, no more than $k - 1$ of the summands can be zero. Thus $u_{12}^* \neq 0$ if there are two or more roots outside the interval $[-1,1]$, a contradiction, and we have shown that at least k of the roots are in the interval $[-1,1]$.

To show that all $k+1$ of the roots are in the interval $[-1,1]$ we continue. We write the polynomial (7) in determinental form

$$P(r) = \frac{(-1)^{k+1}}{\Delta} \begin{vmatrix} 1 & r & \cdots & r^{k+1} \\ f_0 & f_1 & \cdots & f_{k+1} \\ f_1 & f_2 & \cdots & f_{k+2} \\ \cdots & \cdots & \cdots & \cdots \\ f_k & \cdots & \cdots & f_{2k+1} \end{vmatrix} \qquad (13)$$

where

$$\Delta = |X'\Sigma_Y^{-1}X| \quad \text{and} \quad f_L = \sum_i w_i x_i^L \quad L = 0, 1, \ldots, 2k+1$$

This is obtained from Eq. (8) in the following way. Solve for α and write

$$\alpha' = -(X'\Sigma_Y^{-1}X)^{-1}\,\mathbf{f}' = -\frac{\text{ADJ } X'\Sigma_Y^{-1}X}{|X'\Sigma_Y^{-1}X|}\,\mathbf{f}'$$

Let

$$\text{ADJ } X'\Sigma_Y^{-1}X = \begin{bmatrix} A_{11} & -A_{21} & \cdots & (-1)^{k+2}A_{k+1,1} \\ -A_{12} & A_{22} & \cdots & \cdots \\ \cdots & \cdots & \cdots & \cdots \\ (-1)^{k+2}A_{1,k+1} & \cdots & \cdots & A_{k+1,k+1} \end{bmatrix}$$

where A_{ij} is the minor of the (i,j)th element of $X'\Sigma_Y^{-1}X$. We may write the polynomial (7) in the form

$$P(r) = (1,r,\ldots,r^k)\begin{bmatrix} \alpha_0 \\ \cdot \\ \cdot \\ \cdot \\ \alpha_k \end{bmatrix} + r^{k+1}$$

Then

$$P(r) = \frac{-(1,r,\ldots,r^k)}{\Delta} \times \begin{bmatrix} A_{11} & -A_{21} & \cdots & (-1)^{k+2}A_{k+1,1} \\ -A_{12} & A_{22} & \cdots & \cdots \\ \cdots & \cdots & \cdots & \cdots \\ (-1)^{k+2}A_{1,k+1} & \cdots & \cdots & A_{k+1,k+1} \end{bmatrix}\begin{bmatrix} f_{k+1} \\ \cdot \\ \cdot \\ f_{2k+1} \end{bmatrix} + r^{k+1} \quad (14)$$

The coefficient of r^m, $m = 0, \ldots, k$, in Eq. (14) is

$$-\sum_{L=1}^{k+1}(-1)^{m+1}\frac{A_{Lm}}{\Delta}f_{k+L}$$

This is the cofactor of r^m in Eq. (13) expanded along its last column. The coefficient of r^{k+1} is 1 in both equations.

We evaluate $P(r)$ at $r = -1$ and $r = 1$. Let J designate $+1$ or -1. Substituting J into Eq. (13) and subtracting J times the $(m + 1)$st column from the mth column, $m = 1, 2, \ldots, k + 1$, we obtain

$$P(J) = \frac{J^{k+1}}{\Delta} \begin{vmatrix} f_0 - Jf_1 & f_1 - Jf_2 & \cdots & f_k - Jf_{k+1} \\ f_1 - Jf_2 & & \cdots & \cdots \\ \cdots & \cdots & \cdots & \cdots \\ f_k - Jf_{k+1} & \cdots & \cdots & f_{2k} - Jf_{2k+1} \end{vmatrix} \qquad (15)$$

The elements of this determinant have the form $\sum_i w_i x_i^L (1 - Jx_i)$, $L = 0$, $1, \ldots, 2k$. If we denote this determinant by H_j, we see that [by analogy with Eq. (5)]

$$H_J = X'W_J X \qquad (16)$$

where W_J is an $n \times n$ diagonal matrix with elements $w_i(1 - Jx_i)$. Since $\min x_i = -1$ and $\max x_i = 1$, one of the diagonal elements of W_J is always zero; all others are positive. Hence from Eq. (16) it follows that $H_J = X'_J V_J X_J$, where V_J is the $(n - 1) \times (n - 1)$ minor of the zero diagonal element of W_J and X_J is the $(n - 1) \times (k + 1)$ matrix formed from X by striking out the row corresponding to $\min x_i = -1$ for $J = -1$ and corresponding to $\max x_i = 1$ for $J = 1$. Since there are at least $k + 2$ distinct x_i, X_J has rank $k + 1$. V_J is positive definite since it is a diagonal matrix with nonzero nonnegative elements. From this it follows that H_J is positive definite and $|H_J| > 0$. Then, since $|X'\Sigma_Y^{-1}X| > 0$, we can make the following conclusions about $P(J)$ in Eq. (15):

$$\begin{array}{lll} P(1) > 0 & P(\infty) > 0 & \\ P(-1) > 0 & P(-\infty) > 0 & \text{for odd } k \\ P(-1) < 0 & P(-\infty) < 0 & \text{for even } k \end{array} \qquad (17)$$

It was previously shown that k of the roots r_j are in the interval $[-1,1]$. Equation (17) shows that neither -1 nor $+1$ may be roots. Hence there must be k roots in the open interval $(-1,1)$. Furthermore, knowing the sign of $P(r)$ for $r = -1, 1$, and for sufficiently large $|r|$, we reason that all r_j are in the open interval $(-1,1)$, for one exterior root would imply another.

To complete the proof we must show that $X'\Sigma_Y^{-1}X = R'UR$. To do this we define a function u_{gh}, $g, h = 1, \ldots, k+1$ which is proportional to u_{gh}^* if $g \neq h$.

$$u_{gh} = \frac{1}{\prod_{\substack{p=1 \\ p \neq g}}^{k+1}(r_g - r_p) \prod_{\substack{j=1 \\ j \neq h}}^{k+1}(r_h - r_j)} \sum_i w_i \phi_i \prod_{\substack{j=1 \\ j \neq h \\ j \neq g}}^{k+1}(x_i - r_j) \qquad (18)$$

$$u_{gh} = \sum_i w_i \prod_{\substack{p=1 \\ p \neq g}}^{k+1} \frac{x_i - r_p}{r_g - r_p} \prod_{\substack{j=1 \\ j \neq h}}^{k+1} \frac{x_i - r_j}{r_h - r_j} \qquad (19)$$

or

$$u_{gh} = \sum_i w_i F(x_i, r_g) F(x_i, r_h)$$

using the notation of Eq. (1). Form a $(k+1) \times (k+1)$ matrix of these elements u_{gh} and obtain

$$\begin{bmatrix} \sum_i w_i F(x_i, r_1) F(x_i, r_1) & \cdots & \sum_i w_i F(x_i, r_1) F(x_i, r_{k+1}) \\ \vdots & & \vdots \\ \sum_i w_i F(x_i, r_{k+1}) F(x_i, r_1) & \cdots & \sum_i w_i F(x_i, r_{k+1}) F(x_i, r_{k+1}) \end{bmatrix}$$

By analogy with Eq. (5) this matrix is seen to be the product of three matrices:

$$\begin{bmatrix} F(x_1, r_1) & \cdots & F(x_n, r_1) \\ \vdots & & \vdots \\ F(x_1, r_{k+1}) & \cdots & F(x_n, r_k) \end{bmatrix} \begin{bmatrix} w_1 & 0 & 0 & \cdots & 0 \\ 0 & w_2 & & \cdots & 0 \\ \vdots & & & & \vdots \\ 0 & \cdots & & & w_n \end{bmatrix} \begin{bmatrix} F(x_1, r_1) & \cdots & F(x_1, r_{k+1}) \\ \vdots & & \vdots \\ F(x_n, r_1) & \cdots & F(x_n, r_{k+1}) \end{bmatrix}$$

From the preliminary relation (3) we see that this product of matrices can be written in the form

$$(XR^{-1})' \Sigma_Y^{-1} (XR^{-1}) = U \qquad (20)$$

We now show that the matrix U can be considered to be a covariance matrix, i.e., that it is positive definite and symmetric.

The definition (18) of u_{gh} shows that $u_{gh} = 0$ if $g \neq h$. Hence U is a diagonal matrix. The left-hand side of Eq. (20) is nonsingular; thus no diagonal element of U is zero. Let u_n be a typical diagonal element of U; then

$$u_h = \sum_i w_i \prod_{\substack{j=1 \\ j \neq h}}^{k+1} \frac{(x_i - r_j)^2}{(r_j - r)^2}$$

Each $u_h > 0$, and we have shown the U is a positive definite symmetric matrix.

We can rewrite Eq. (20) as $X'\Sigma_Y^{-1}X = R'UR$ since R^{-1} is nonsingular and has an inverse. Since the (1,1) element of $X'\Sigma_Y^{-1}X$ is the total information $Q = \sum_i w_i$, and since the (1,1) element of $R'UR$ is $\sum_{h=1}^{k+1} u_h$, it follows that $Q = \sum_{h=1}^{k+1} u_h$. Thus we interpret $R'UR$ as the information matrix of n observations made at $k+1$ distinct points r_j. We have proved the main result of this section.

EXAMPLE

Let $k = 2$, and permit N independent observations $y(x_t)$, $t = 1$, ..., N, of equal variance σ^2 to be taken in the specified interval $x_L \leq x_t \leq x_H$. Let $Y(\xi)$ be the least-squares estimator of $P(\xi)$. The problem is to find the spacing of the N observations that will minimize the maximum variance of $Y(\xi)$ for $x_L \leq \xi \leq x_H$.

The variance of $Y(\xi)$ is $\sigma^2_{Y(\xi)} = \xi(X'WX)^{-1}\xi'$. Whatever the optimum spacing may be, it will give rise to some matrix $X'WX$; let this be $(X'WX)_0$. From the given results, it follows that there exists a matrix $R'UR = (X'WX)_0$. Since $m = 2$,

$$R = \begin{bmatrix} 1 & r_1 & r_1^2 \\ 1 & r_2 & r_2^2 \\ 1 & r_3 & r_3^2 \end{bmatrix} \quad \text{and} \quad U = \begin{bmatrix} n_1 & 0 & 0 \\ 0 & n_2 & 0 \\ 0 & 0 & n_3 \end{bmatrix} \frac{1}{\sigma^2}$$

implying n_j observations at r_j satisfying $x_L \leq r_j \leq x_H$ and $\Sigma n_j = N$, $j = 1, 2, 3$.

Let $\bar{y}_j$ be the average of the n_j observations at r_j; let **n** be the column vector $[\bar{y}_1 \; \bar{y}_2 \; \bar{y}_3]'$; and let **a** be the column satisfying the normal equations $R'UR\mathbf{a} = R'U\mathbf{n}$. Since R and U are nonsingular, $R\mathbf{a} = \mathbf{n}$. Since $R\mathbf{a}$ is the column vector $[Y(r_1) \; Y(r_2) \; Y(r_3)]'$, $Y(\xi)$ passes through $\bar{y}_i$ at

$\xi = r_j$. Hence, $Y(\xi)$ may be written in the Lagrange form

$$Y(\xi) = \frac{(\xi - r_2)(\xi - r_3)}{(r_1 - r_2)(r_1 - r_3)} \bar{y}_1 + \frac{(\xi - r_1)(\xi - r_3)}{(r_2 - r_1)(r_2 - r_3)} \bar{y}_2 + \frac{(\xi - r_1)(\xi - r_2)}{(r_3 - r_1)(r_3 - r_2)} \bar{y}_3 \qquad (21)$$

It follows that $\sigma^2_{Y(r_j)} = \sigma^2/n_j$. For any such spacing, let $\sigma^2_{\max}$ be the maximum variance of $Y(\xi)$ in the interval. Then

$$\frac{\sigma^2}{n_j} \leq \sigma^2_{\max}$$

and thus

$$\frac{\sigma^2}{3} \sum_1^3 \frac{1}{n_j} \leq \sigma^2_{\max} \qquad (22)$$

The minimum value of $\sum_1^3 \frac{1}{n_j}$ constrained by $\sum_1^3 n_j = N$ is found to be $9/N$ with $n_j = N/3$. Hence, from (22)

$$\frac{3\sigma^2}{N} \leq \min \sigma^2_{\max} \qquad (23)$$

Note from (21) that $\sigma^2_{Y(\xi)}$ increases as ξ departs from the smallest and the largest r_j in direction of leaving the interval x_L, x_H. Hence, locate $N/3$ observations at x_L and x_H. Note that $\sigma^2_{Y(\xi)}$ has one differentiable maximum occurring in the interior of the interval. Locate $N/3$ observations at $(x_L + x_H)/2$. From symmetry, the differentiable maximum then occurs at $(x_L + x_H)/2$. Hence the maximum $\sigma^2_{Y(\xi)}$ for $x_L \leq \xi \leq x_H$ for the spacing $r_1 = x_L, r_2 = (x_L + x_H)/2, r_3 = x_H, n_j = N/3$ is $3\sigma^2/N$. Inequality (23) assures that this particular spacing gives the desired minimax variance and that this variance is $3\sigma^2/N$, for the equality has been produced.

3 □ THE OPTIMAL LOCATION OF THE REGRESSOR VARIABLES

The Gauss-Markoff theorem proves that the Markoff estimators are minimum-variance-unbiased linear estimators of the regression parameters p. The generalized variance (GV) of these estimators for a set of observa-

tions taken at n distinct locations x_i, $i = 1, \ldots, n$, is given by the formula
$$GV = |X'\Sigma_Y^{-1}X|^{-1}$$
By the result of De la Garza we have
$$GV = |R'UR|^{-1} \tag{1}$$

Let n_j observations be made at locations r_j, $j = 1, \ldots, k+1$, and let $\sum_{j=1}^{k+1} n_j = n$. If we assume that $\text{var } y_i = \sigma^2$ for all i, then u has the form

$$\begin{bmatrix} \frac{n_1}{\sigma^2} & 0 & \cdots & 0 \\ 0 & \frac{n_2}{\sigma^2} & \cdots & 0 \\ \cdot & \cdot & \cdot & \cdot \\ 0 & 0 & \cdots & \frac{n_{k+1}}{\sigma^2} \end{bmatrix}$$

We would like to determine r_j so as to minimize the generalized variance given by Eq. (1). We may write (1) in the following form

$$\frac{1}{GV} = \left| \begin{bmatrix} 1 & 1 & \cdots & 1 \\ r_1 & r_2 & \cdots & r_k \\ \cdot & \cdot & \cdot & \cdot \\ r_1^k & r_2^k & \cdots & r_{k+1}^k \end{bmatrix} \begin{bmatrix} \frac{n_1}{\sigma^2} & 0 & \cdots & 0 \\ 0 & \frac{n_2}{\sigma^2} & \cdots & 0 \\ \cdot & \cdot & \cdot & \cdot \\ 0 & 0 & \cdots & \frac{n_{k+1}}{\sigma^2} \end{bmatrix} \begin{bmatrix} 1 & r_1 & \cdots & r_1^k \\ 1 & r_2 & \cdots & r_2^k \\ \cdot & \cdot & \cdot & \cdot \\ 1 & r_{k+1} & \cdots & r_{k+1}^k \end{bmatrix} \right|$$

$$= \begin{vmatrix} 1 & 1 & \cdots & 1 \\ r_1 & r_2 & \cdots & r_{k+1} \\ \cdot & \cdot & \cdot & \cdot \\ r_1^k & r_2^k & \cdots & r_{k+1}^k \end{vmatrix}^2 \frac{\prod_{j=1}^{k+1} n_j}{\sigma^{2k+2}} \tag{2}$$

This is a Vandermonde determinant with value $\prod_{i<j}^{k+1} (r_i - r_j)^2$; consequently,

$$\frac{1}{GV} = \frac{1}{\sigma^{2k+2}} \prod_{i<j}^{k+1} (r_i - r_j)^2 \prod_{j=1}^{k+1} n_j \tag{3}$$

48 PARAMETRIC ESTIMATION

The minimization of the generalized variance is equivalent to the maximization of Eq. (3), which is accomplished in the following two lemmas.

Lemma 1

The maximum value of $\prod_{j=1}^{k+1} n_j$ subject to the constraint $\sum_{j=1}^{k+1} n_j = n$ occurs when $n_l = n_j$, $j, l = 1, \ldots, k+1$.

Proof:
Introduce the Lagrange multiplier ρ and maximize the expression

$$\prod_{j=1}^{k+1} n_j + \rho \left(\sum_{j=1}^{k+1} n_j - n \right)$$

We take partial derivatives with respect to n_j and obtain for all j

$$\rho = - \prod_{i=1}^{k+1} \frac{n_i}{n_j}$$

Hence

$$\prod_{i=1}^{k+1} \frac{n_i}{n_j} = \prod_{i=1}^{k+1} \frac{n_i}{n_l} \quad \text{for all } i \text{ and } j$$

Thus the maximum occurs when $n_l = n_j$

Lemma 2

Let $\Delta = \prod_{i<j}^{k+1} (r_i - r_j)^2$, $-1 \leq r_j \leq 1$. Then Δ reaches a maximum value only for the zeros of the function

$$F_{k+1} = \int_r^1 p_k(t) \, dt \qquad (4)$$

where $p_k(t)$ is the kth Legendre polynomial.

Proof:
This proof is due to Schur and Stieltjes. Since Δ is a continuous function of the r_j on the closed interval, there does exist a maximum value of Δ. At such a maximum value the roots must all be different, for other-

wise $\Delta = 0$. Since Δ is symmetric in the r_j, we can assume
$$-1 \leq r_1 < r_2 < \cdots < r_{k+1} \leq 1$$

For $k + 1 = 2$, $\Delta \leq 4$ and $r_1 = -1$ and $r_2 = 1$. For $k + 1 > 2$ we shall obtain first the conditions on r_1 and r_{k+1} for a maximum, then the conditions on r_p, $1 < p < k + 1$. The only factors in Δ which depend upon r_1 and r_{k+1} are

$$\prod_{j=1}^{k} (r_{k+1} - r_j)^2 \prod_{i=2}^{k} (r_i - r_1)^2$$

This factor obviously reaches an end-point maximum when $r_1 = -1$ and $r_{k+1} = 1$. Note that $1 < p < k + 1$ now implies that $-1 < r_p < 1$.

For $1 < p < k + 1$, the condition for a maximum of Δ is

$$\frac{1}{\Delta} \frac{\partial}{\partial r_p} \Delta = 2 \sum_{\substack{j=1 \\ j \neq p}}^{k+1} \frac{1}{r_p - r_j} = 0 \tag{5}$$

We define
$$f(r) = (r - r_1) \cdots (r - r_{k+1})$$
and
$$f_p(r) = \frac{f(r)}{r - r_p}$$

Now,
$$f'_p(r) = \sum_{\substack{j=1 \\ j \neq p}}^{k+1} \frac{f_p(r)}{r - r_j} \quad \text{and} \quad f'(r) = \sum_{j=1}^{k+1} \frac{f(r)}{r - r_j}$$

where the prime denotes a derivative with respect to r. Hence

$$\frac{f'_p(r)}{f_p(r)} = \sum_{\substack{j=1 \\ j \neq p}}^{k+1} \frac{1}{r - r_j}$$

and from Eq. (5) we obtain

$$\frac{f'_p(r_p)}{f_p(r_p)} = 0 \quad \text{and} \quad f_p(r_p) \neq 0 \tag{6}$$

Further,
$$f(r) = (r - r_p) f_p(r)$$
$$f'(r) = (r - r_p) f'_p(r) + f_p(r)$$
$$f''(r) = (r - r_p) f''_p(r) + 2 f'_p(r)$$

Hence at $r = r_p$, from Eq. (6),
$$f''(r_p) = 2 f'_p(r_p) = 0$$

This shows that $f''(r) = 0$ has the same roots as $f(r) = 0$ on the open interval $(-1,1)$. Since $f(r) = 0$ has two more roots than $f''(r) = 0$, we conclude that -1 and 1 are roots of $f(r) = 0$. Consequently, we see that $(r^2 - 1)f''(r) = 0$ has the same roots as $f(r) = 0$. Hence,

$$(r^2 - 1)f''(r) = cf(r);$$

by equating the coefficient of r^{k+1} on both sides, we see the constant is $k(k + 1)$. Thus

$$(r^2 - 1)f''(r) = k(k + 1)f(r) \tag{7}$$

Now Eq. (7) has as its solution the function of the Legendre polynomial given in Eq. (4). Thus it will be verified by direct substitution of (4) into (7), to obtain

$$(r^2 - 1) \frac{d^2}{dr^2} \int_r^1 p_k(t) \, dt = k(k + 1) \int_r^1 p_k(t) \, dt$$

We may write Legendre's differential equation in the form

$$\frac{d}{dt} (1 - t^2) p_k'(t) = -k(k + 1) p_k(t)$$

If we integrate both sides of this equation for $t = r$ to $t = 1$, we obtain

$$(1 - r^2) p_k'(r) = k(k + 1) \int_r^1 p_k(t) \, dt$$

Obviously

$$(1 - r^2) p_k'(r) = (r^2 - 1) \frac{d^2}{d^2 r} \int_r^1 p_k(t) \, dt$$

and the lemma is proved.

Thus we have obtained the optimal locations of the regressor variables on the interval $[-1,1]$; they are listed for given polynomial degrees in Prevost [16].

We now give a solution due to Guest [5], the uniqueness of which is proved by Antle [1]. This will throw light on the structure of the solution.

4 □ SPACING FOR MINIMAX VARIANCE

Assume the setup of the given previous section. Then the fitted value is given by

$$u_k(x) = \sum_{j=0}^{k} L_j(x) \bar{y}_j \tag{1}$$

where $L_j(x)$ is the lagrangian coefficient corresponding to the point of observation x_j and $\bar{y}_j$ is the mean of the observed values at this point. The variance of the fitted value is var $u_k(x) = \sum_{j=0}^{k} L_j^2(x)$ var $\bar{y}_j$.

At a point of observation

$$L_j(x_p) = \delta_{jp}$$

where

$$\delta_{jp} = \begin{cases} 1 & j = p \\ 0 & j \neq p \end{cases}$$

and

$$\text{var } u_k(x_j) = \text{var } \bar{y}_j$$

The largest value of this variance will be as small as possible when the n observations are equally divided among the $k+1$ points. When this is done,

$$\text{var } u_k(x_j) = \frac{(k+1)\sigma^2}{n} \tag{2}$$

$$\text{var } u_k(x) = \sum_{j=0}^{k} L_j^2(x) \frac{(k+1)\sigma^2}{n} \tag{2a}$$

Since this is a polynomial of degree $2k$, the minimax-variance conditions are obtained when the maxima of var $u_k(x)$ are at the $k-1$ internal points x_j and the end points x_0 and x_k are $+1$ and -1; for the var $u_k(x)$ never exceeds

$$\frac{(k+1)\sigma^2}{n}$$

in the range $+1$ to -1. The minimax-variance conditions are thus

$$L_j'(x_j) = 0 \quad j = 1, \ldots, k-1 \tag{3}$$

Now, if

$$F(x) = \prod_{j=0}^{k} (x - x_j) \tag{4}$$

then

$$L_j(x) = \frac{F(x)}{(x - x_j)F'(x_j)}$$

and so

$$F''(x) = [(x - x_j)L_j''(x) + 2L_j'(x)]F'(x_j)$$

and (3) is equivalent to

$$F''(x_j) = 0 \quad j = 1, \ldots, k-1 \tag{5}$$

The function $F(x)$ will be of the form $\alpha(x^2 - 1)\phi_{p-1}(x)$, where the polynomial $\phi_{k-1}(x)$ of degree $k-1$ is determined by the $p-1$ equations (5). The polynomial which satisfies these equations is readily shown to be the derivative $p'_k(x)$ of the Legendre polynomial. For if

$$F(x) = \alpha(x^2 - 1)p'_k(x) \tag{6}$$

then

$$F'(x) = \alpha \frac{d}{dx}(x^2 - 1)p'_k(x) = \alpha k(k+1)p_k(x)$$

and

$$F''(x) = \alpha k(k+1)p'_k(x)$$

and so $F''(x)$ vanishes at the internal points $F(x) = 0$.

The points of observation for minimax variance are then to be located at $+1$, -1, and the roots of $p'_k(x) = 0$.

Since the internal points of observation are points of maximum variance, the variance will be given by an equation of the form

$$\mathrm{var}\, u_k(x) = [1 + \beta(x^2 - 1)p'^2_k(x)] \frac{(k+1)\sigma^2}{n} \tag{7}$$

The minima of the variance curve then occur at points for which

$$xp'_k(x) + (x^2 - 1)p''_k(x) = 0$$

and this equation is equivalent to

$$xp'_k(x) = k(k+1)p_k(x) \tag{8}$$

From (2a)

$$\mathrm{var}\, u_k(x) = \sum_{j=0}^{k} \left[\frac{\alpha(x^2 - 1)p'_k(x)}{(x - x_j)\alpha k(k+1)p_k(x_j)} \right]^2 \frac{(k+1)\sigma^2}{n} \tag{9}$$

and so, on comparing the coefficients of $x^2 p'^2_k(x)$ in (7) and (9),

$$\beta = \sum_{j=0}^{k} [k(k+1)p_k(x_j)]^{-2}$$

The Lobatto quadrature formula (Hilderbrand [6, pp. 319–345]) with $f(x) \equiv 1$ gives

$$\int_{-1}^{1} dx = \sum_{j=0}^{n} 2[k(k+1)p_k^2(x_j)]^{-1} = 2$$

Thus the explicit formula for the variance of the fitted value is

$$\operatorname{var} u_k(x) = \left[1 + \frac{x^2 - 1}{k(k+1)} p_k'^2(x)\right] \frac{(k+1)\sigma^2}{n} \tag{10}$$

In the region of extrapolation, when $|x|$ is large,

$$p_k'(x) \doteq k \frac{(2k)!}{2^k k_1^2} x^{k-1}$$

$$\operatorname{var} u_k(x) \doteq k \left[\frac{(2k)!}{2^k k^2}\right]^2 x^{2k} \frac{\sigma^2}{r}$$

PROBLEMS

1. (Smith) Let $k = 3, 4, 5, 6$ (consider a polynomial of fixed degree k), and permit N independent observations $Y(X_i)\ i = 1, 2, \ldots, N$, of equal variance σ^2 to be taken in the specified interval $x_L \leq x_i \leq x_H$. Let $Y(x)$ be the least-squares estimator of polynomial $p(x)$. Find the spacing of the N observations that will minimize the maximum variance of $Y(x)$ for $x_L \leq x \leq x_H$.

2. (Antle) Prove that the solution given in Sec. 4 is unique.

3. Prove or disprove the De la Garza theorem for the following functions (assume all other conditions of the theorem):

$$p_1(x) = e^{-x}(a_0 + a_1 x + \cdots + a_k x^k)$$
$$p_2(x) = e^{-x^2}(a_0 + a_1 x + \cdots + a_k x^k)$$

4. (Hoel and Levine) Let $-1 \leq x_i \leq 1,\ i = 1, 2, \ldots, n$, denote the selected values of a variable x at which observations are to be made on a related variable y corresponding to those selected values. Let $y(x_i)$ denote the observed value of Y corresponding to x_i. Assume that $y(x_i)$, $i = 1, \ldots, n$, are uncorrelated random variables with a common variance σ^2. Let the means of y's lie on a polynomial curve of known degree k, that is,

$$E[Y(x_i)] = a_0 + a_1 x_i + \cdots + a_k x_i^k$$

Let $k + 1$ points be chosen for estimating a polynomial of degree k.

(a) Show that the least-squares estimator is given by

$$\hat{y}(x) = \sum_{i=1}^{k} L_i(x)\bar{y}_i$$

where $\bar{y}$ is mean of the observations taken at x_i and

$$L_i(x) = \frac{(x - x_0) \cdots (x - x_{i-1})(x - x_{i+1}) \cdots (x - x_k)}{(x_i - x_0) \cdots (x_i - x_{i+1})(x_i - x_{i+1}) \cdots (x_i - x_k)}$$

and

$$\text{var } \hat{y}(x) = \frac{\sigma^2}{n} \sum_{i=0}^{k} \frac{L_i^2(x)}{p_i}$$

where p_i is proportion of the observations at x_i.

(b) Prove the following. If the p_i, $i = 0, 1, \ldots, k$, are allowed to vary continuously in $(0,1)$ under the restriction $\sum_0^k p_i = 1$, then for a fixed $x \neq x_i$, $i = 0, 1, \ldots, k$ the choice

$$p_i = \frac{|L_i(x)|}{\Sigma |L_i(x)|} \qquad i = 0, 1, \ldots, k \tag{1}$$

will minimize var $\hat{y}(x)$.

(c) Let $x_0 < x_1 < \cdots < x_k$ be $k + 1$ distinct points in $[-1,1]$, and let $G(x)$ be a kth-degree polynomial possessing the property that $G(x_0) = 0$ and the property that $G(x_i) = (-1)^i$, $i = 1, \ldots, k$. Then prove that $G(x)$ possesses at least j roots, counting multiplications, in the closed interval $[x_0, x_j]$.

5. (Hoel and Levine) Assume the conditions of Prob. 4, and if the minimizing p's given by (1) are used, prove that the $k + 1$ observation points that will minimize var $\hat{y}(x)$ at $x > 1$ are given by the Tchebychev points

$$x_i = -\cos\frac{i\pi}{k} \qquad i = 0, 1, 2, \ldots, k$$

6. (Hoel) Assume the setup of Sec. 3 and prove the following fact. The generalized variance of the ordinates of a polynomial regression curve at $k + 1$ arbitrary points in an interval will be minimized when the generalized variance of the estimators of the regression coefficients is minimized.

REFERENCES

1 ANTLE, C. E.: The Uniqueness of the Spacing of Observations in Polynomial Regression for Minimax Variance of Fitted Values, *Ann. Math. Statist.*, vol. 33, pp. 810–811, 1962.

2 DAVID, H. A., and B. E. ARENS: Optimal Spacing in Regression Analysis, *Ann. Math. Statist.*, vol. 30, pp. 1072–1081, 1959.

3 DE LA GARZA, A.: Spacing of Information in Polynomial Estimation, *Ann. Math. Statist.*, vol. 25, pp. 123–130, 1954.

4 ELFVING, G.: Optimum Allocation in Linear Regression Theory, *Ann. Math. Statist.*, vol. 23, pp. 204–262, 1952.

5 GUEST, P. G.: The Spacing of Observations in Polynomial Regression, *Ann. Math. Statist.*, vol. 29, pp. 294–299, 1958.

6 HILDERBRAND, F. B.: "Methods of Applied Mathematics," Prentice-Hall, Inc., Englewood Cliffs, N.J., 1952.

7 HOEL, P. G.: Efficiency Problems in Polynomial Estimation, *Ann. Math. Statist.*, vol. 29, pp. 1134–1145, 1958.

8 HOEL, P. G.: Some Properties of Optimal Spacing in Polynomial Estimation, *Ann. Inst. Statist. Math.*, vol. 13, no. 1, pp. 1–8, 1961.

9 HOEL, P. G.: Minimax Designs in Two Dimensional Regression, *Ann. Math. Statist.*, vol. 36, pp. 1097–1106, 1965.

10 HOEL, P. G.: Optimum Designs for Polynomial Extrapolation, *Ann. Math. Statist.*, vol. 36, pp. 1483–1493, 1965.

11 HOEL, P. G., and A. LEVINE: Optimal Spacing and Weighting in Polynomial Prediction, *Ann. Math. Statist.*, vol. 35, pp. 1553–1560, 1964.

12 KENDALL, M. G., and A. STUART: "The Advanced Theory of Statistics," vol. 2, Charles Griffin & Company, Ltd., London, 1961.

13 Kiefer, J.: Optimum Experimental Designs, *J. Roy. Statist. Soc. ser. B*, vol. 21, pp. 272–319, 1959.

14 KIEFER, J., and J. WOLFOWITZ: Optimum Designs in Regression Problems, *Ann. Math. Statist.*, vol. 30, pp. 272–294, 1959.

15 KIEFER, J., and J. WOLFOWITZ: On a Theorem of Hoel and Levine on Extrapolation, *Ann. Math. Statist.*, vol. 36, pp. 1627–1655, 1965.

16 PREVOST, G.: "Tables de fonctions sphériques," Gauthier-Villiers, Paris, 1933.

17 SCHEFFÉ, H.: "The Analysis of Variance," John Wiley & Sons, Inc., New York, 1959.

18 SCHUR, I.: About the Distribution of Roots of Certain Algebraic Equations with Integral Coefficients, *Math. Z.*, vol. 1, pp. 377–402, 1918.

19 SHOHAT, J. A., and J. D. TAMARKIN: The Problems of Moments, *Math. Survey*, no. 1, 1943.

20 SMITH, K.: On the Standard Deviations of Adjusted and Interpolated Values of an Observed Polynomial Function, etc., *Biometrika*, vol. 12, pp. 1–85, 1918.

chapter Four

Completeness

1 □ INTRODUCTION

In this chapter we shall investigate a property, namely, completeness, of measurable functions of random variables for a certain class of probability measures. For our whole space we choose R^x, the real line, or a subset of the real line, consisting of points x, and we shall consider certain fixed countably additive families F^x of subsets of R^x. A nonnegative set function p^x defined for all $A \in F^x$ is said to be a measure on F^x if it is countably additive, i.e., if $P^x(A_1 \cup A_2) = P^x(A_1) + P^x(A_2)$ for disjoint sets A_1 and A_2 in F^x; it is a probability measure if in addition $P^x(R^x) = 1$. A family of probability measures indexed by a subscript θ belonging to some abstract space ω will be denoted by $P^x = \{P_\theta^x | \theta \in \omega\}$. For example, $P^x = \{N(\theta,1) | -\infty < \theta < \infty\}$ is the family of normal probability measures with variance 1. A set $N \in F^x$ for which $P_\theta^x(N) = 0$ will be called a *null set* for the measure P_θ^x. If a statement about the points of R^x is true for all x in $R^x - N$, we shall say it is valid almost everywhere P_θ^x. We shall say it is valid almost everywhere P^x if N is a null set for P^x, that is, for every measure P_θ^x in the family P^x. We shall abbreviate this by writing a.e. P_θ^x or a.e. P^x after the statement. We give notation for two dimensions, which can easily be extended to higher dimensions. Let $x = (x',x'')$, $R^x = R^{x'} \times R^{x''}$, $F^x = F^{x'} \times F^{x''}$, $\theta = (\theta',\theta'')$, $\omega = \omega' \times \omega''$, $\mathcal{L}^\theta = \mathcal{L}^{\theta'} \times \mathcal{L}^{\theta''}$, where $\mathcal{L}^{\theta'}$ and $\mathcal{L}^{\theta''}$ are σ-fields generated by Borel sets of ω' and ω'', respectively, $L^\theta = L^{\theta'} \times L^{\theta''}$, where L is a Lebesgue measure.

Given a family $P^x = \{P_\theta^x | \theta \in \omega\}$ of probability measures, consider integrals of the form

$$\int_{R^x} f(x) \, dP_\theta^x \qquad (1)$$

where $f(x)$ is real-valued and measurable (F^x). The value (if any) of this integral will in general depend on which measure P_θ^x of the family is used. The integral in (1) may be regarded as transforming the function $f(x)$ defined on R^x to a function of θ defined on some subset of ω. Under this transformation the function that is everywhere zero on R^x goes into the function that is everywhere zero on ω. Thus completeness means roughly that the zero function on R^x is the only function which goes into the zero function on ω; it is a unicity property of the transform.

In Sec. 2 we define complete families of probability measures and complete statistics, give illustrative examples, and discuss some pertinent properties. In Sec. 3 we define boundedly complete families of probability measures and show that completeness implies boundedly completeness but that the converse is not true. Then in Sec. 4 we discuss strongly complete families of probability measures and indicate that strong completeness implies completeness. We then investigate properties of strongly complete families of measures. In the subsequent chapters we shall make use of these properties in order to establish some important properties of estimates.

2 □ COMPLETE FAMILIES OF PROBABILITY MEASURES

Definition 1

The family P^x of measures is complete if

$$\int_{R^x} f(x) \, dP_\theta^x = 0 \qquad \text{for all } \theta \in \omega \qquad (1)$$

implies that
$$f(x) = 0 \quad \text{a.e. } P^x$$

Definition 2

A statistic t is said to be complete if the probability measure induced by t is complete over the space $(\mathfrak{J},\beta)$ for which $\mathfrak{J}$ is the whole space of values of t and β is the σ-field generated by Borel sets of $\mathfrak{J}$.

Theorem 1

If t is complete for $\{p_\theta{}^x | \theta \in \omega\}$, then any statistic t^* depending on x only through t is also complete for $\{p_\theta{}^x | \theta \in \omega\}$.

Proof:

Let $t^* = g(t)$ for some g.

Suppose there is a function h satisfying the following conditions:

(i) $h(t^*)$ is measurable.
(ii) $E_\theta[h(t^*)] = 0$ for all $\theta \in \omega$.

Thus $h[g(t)]$ is measurable, and $E_\theta\{h[g(t)]\} = 0$, $\theta \in \omega$, implies that the function $h[g(t)]$ is zero by the completeness property of t. Thus $h(t^*) = 0$, a.e. P^x, which is the required result.

We give now some illustrative examples.

EXAMPLE 1

Let $p^x = \{b(\theta,n) | 0 \leq \theta \leq 1\}$ be the family of binomial distributions corresponding to n independent trials with constant probability θ, $0 \leq \theta \leq 1$; $p_\theta{}^x$ assigns the probability

$$\binom{n}{x} \theta^x (1 - \theta)^{n-x} \tag{2}$$

to the points $x = 0, 1, \ldots, n$ and zero probability elsewhere. Then P^x is complete.

Proof:

We have to show that

$$\sum_{x=0}^{n} f(x) \binom{n}{x} \theta^x (1 - \theta)^{n-x} \equiv 0 \tag{3}$$

implies $f(x) = 0$ for $x = 0, 1, \ldots, n$. We can write (3) in the form

$$\sum_{x=0}^{n} a(x)\varphi^x \equiv 0 \qquad (4)$$

where

$$a(x) = f(x)\binom{n}{x} \quad \text{and} \quad \frac{\theta}{1-\theta} = \varphi$$

Equation (4) is a polynomial in φ of degree n and vanishes for every value of φ, which is possible only when $a(x) = 0$ for $x = 0, 1, 2, \ldots, n$. Hence, $f(x) = 0$ for $x = 0, 1, 2, \ldots, n$. That is, $f(x) = 0$, a.e. p^x.

EXAMPLE 2

Let p^x be the family of Poisson distributions, so that $p_\theta{}^x$ assigns the measure $e^{-\theta}\theta^x/x!$ to the set of point $x = 0, 1, 2, \ldots$ and measure zero to the complement of this set. Then p^x is complete.

Proof:

We have to show that

$$e^{-\theta}\sum_{x=0}^{\infty} f(x)\frac{\theta^x}{x!} = 0 \qquad \text{for } \theta > 0 \qquad (5)$$

implies that $f(x) = 0$ for $x = 0, 1, 2, \ldots$.

Since $f(x)$ is the coefficient of θ^x in the power-series expansion of zero, it follows that $f(x) = 0$, a.e. p^x.

EXAMPLE 3

Let p^x be the family of normal distributions, so that $p_\theta{}^x$ assigns the measure

$$\frac{1}{\sqrt{2\pi}} e^{-(x-\theta)^2/2}\, dx$$

to the interval dx for $-\infty < x < \infty$. Then p^x is complete.

Proof:

We have to prove that

$$\frac{1}{\sqrt{2\pi}} \int_{-\infty}^{\infty} f(x) e^{-(x-\theta)^2/2}\, dx = 0 \qquad (6)$$

implies $f(x) = 0$, a.e. p^x.

By removing the nonzero constant factor, we have

$$\int_{-\infty}^{\infty} f(x) e^{-x^2/2} e^{x\theta} \, dx = 0$$

Hence the Laplace transform of the function

$$f(x) e^{-x^2/2}$$

is zero identically. But the transform of the function 0 is also zero identically. Hence, by the uniqueness property of Laplace transforms, it follows that

$$f(x) e^{-x^2/2} = 0$$

and hence $f(x) = 0$, a.e. Lebesgue measure. Hence p^x is a complete family of distributions.

EXAMPLE 4

Let $X_1, X_2, \ldots, X_n$ be independent random variables with common density function

$$p_N(X_i = k) = \frac{1}{N} \qquad k = 1, 2, \ldots, N \qquad (7)$$

where $N \in \omega$, the set of positive integers. Then the family of distributions induced by $t = \max_{1 \le i \le n} X_i$ is complete.

Proof:

One can easily find that the density function of t is

$$p_N(t = y) = \frac{y^n}{N^n} - \frac{(y-1)^n}{N^n} \qquad (8)$$

Now we have to show that

$$\frac{1}{N^n} \sum_{y=1}^{N} f(y)[y^n - (y-1)^n] = 0 \qquad \text{for } N = 1, 2, \ldots \qquad (9)$$

implies that $f(y) = 0$ for $y = 1, 2, \ldots, N$ for

$$\begin{array}{ll} N = 1 & f(1) = 0 \\ N = 2 & f(1) = 0 \qquad f(2) = 0 \end{array}$$

and so on. Hence the family of distributions induced by t is complete.

We give an example in well-known terminology.

EXAMPLE 5

Let X be a random variable with density

$$f_N(x) = \begin{cases} (N+1)x^N & 0 \le x \le 1 \\ 0 & \text{otherwise} \end{cases} \qquad (10)$$

If $N \in \omega$, the set of nonnegative integers, then $t(x) = x$ defines a complete statistic. That is, if $h(x)$ is a measurable function on the interval $[0,1]$ such that

$$E_{f_n}(h) = \int_0^1 h(x)\,(N+1)x^N\,dx = 0 \qquad N = 0, 1, 2, \ldots \qquad (11)$$

then $h(x) = 0$ for almost all x.

Proof:

Let

$$H(x) = \int_0^x h(s)\,ds \qquad 0 \le x \le 1 \qquad (12)$$

Then

$$\int_0^1 x^N\,dH(x) = 0 \qquad N = 0, 1, 2, \ldots \qquad (13)$$

but

$$\int_0^1 x^N\,dH(x) = \int_0^1 H(x)\,dx^N + H(1)\cdot 1 - H(0)\cdot 0 \qquad (14)$$

Now $H(1) = E_{f_0}(h) = 0$, and so

$$\begin{aligned} 0 &= \int_0^1 H(x)\,d(x^N) \qquad N = 1, 2, \ldots \\ &= N\int_0^1 H(x)x^{N-1}\,dx \end{aligned} \qquad (15)$$

Therefore

$$\int_0^1 x^N H(x)\,dx = 0 \qquad N = 0, 1, 2, \ldots \qquad (16)$$

Let

$$K(x) = \int_0^x H(s)\,ds \qquad (17)$$

Proceeding as above, we determine that

$$\int_0^1 x^N K(x)\,dx = 0 \qquad N = 0, 1, 2, \ldots \qquad (18)$$

Since $H(s)$ is the difference between two monotone functions, K is continuous on $[0,1]$. Clearly for any polynomial

$$p(x) = a_0 x^r + a_1 x^{r-1} + \cdots + a_r$$

$$\int_0^1 p(x) K(x)\,dx = 0$$

Since $K(x)$ is continuous on $[0,1]$, for any $\varepsilon > 0$ there is a polynomial $p(x)$ such that $|p(x) - K(x)| \leq \varepsilon$ for x in $[0,1]$.

Now
$$\left| \int_0^1 K^2(x)\, dx \right| = \left| \int_0^1 K(x)[K(x) - p(x)]\, dx \right|$$
$$\leq \int_0^1 |K(x)|\, |K(x) - p(x)|\, dx$$
$$\leq \varepsilon \int_0^1 |K(x)|\, dx < \infty$$

since $K(x)$ is continuous on the closed interval $[0,1]$. Since ε is arbitrary,
$$\int_0^1 K^2(x)\, dx = 0 \tag{19}$$

and so $K^2(x) = K(x) = 0$ for almost all x. Since
$$K(x) = \int_0^x H(s)\, ds = 0 \quad \text{a.e.}$$
it follows that
$$H(x) = \int_0^x h(s)\, ds = 0 \quad \text{a.e.}$$
and finally
$$h(x) = 0 \quad \text{a.e.}$$

EXAMPLE 6

Let X be a random variable with density function
$$f_\lambda(x) = \begin{cases} \lambda e^{-\lambda x} & x > 0 \\ & \lambda \in \omega, \text{ the set of positive integers} \\ 0 & x \leq 0 \end{cases}$$

Proof:

We shall show that X is complete.

Let $\gamma = e^{-x}$. Then
$$g_\lambda(\gamma) = \begin{cases} \lambda \gamma^{\lambda - 1} & 0 \leq \gamma \leq 1 \\ 0 & \text{otherwise} \end{cases}$$

where g_λ is the density function of γ. By the previous example, we know that γ is complete for λ in the set of positive integers.

Since $x = -\log \gamma$, x is completely determined by γ and so, using Theorem 1, we conclude that X is complete since γ is.

Now we shall give an example of a family of probability measures which is not complete.

EXAMPLE 7

Let X be a random variable with uniform density function over an interval $[\theta, \theta + 1]$, where θ is any real number. The family of probability measures so determined is not complete. We construct a counterexample as follows.

Define

$$h(x) = \begin{cases} 1 & n < x \leq \tfrac{1}{2} + n \\ -1 & \tfrac{1}{2} + n < x < n + 1 \end{cases}$$

for every integer n. Then if $\theta < n \leq \theta + \tfrac{1}{2}$,

$$\int_\theta^{\theta+1} h(x)\,dx = \int_\theta^n h(x)\,dx + \int_n^{n+\tfrac{1}{2}} h(x)\,dx + \int_{n+\tfrac{1}{2}}^{\theta+1} h(x)\,dx$$
$$= -(n - \theta) + \tfrac{1}{2} - (\theta + 1 - n - \tfrac{1}{2}) = 0$$

The integral can be shown to be zero again for $\theta + \tfrac{1}{2} < n \leq \theta + 1$. Thus we have a function h such that $E_\theta[h(X)] = 0$ for all θ and which is nowhere zero. Hence this family of probability measure is not complete.

We discuss the completeness of $\mathfrak{X}^2$-distribution and the noncompleteness of Cauchy distribution in the following examples.

EXAMPLE 8

Let X be a random variable with density function

$$f(x,n) = \begin{cases} \dfrac{1}{\Gamma n\, 2^n} x^{n-1} e^{-x/2} & x > 0 \\ 0 & x \leq 0 \end{cases} \quad \text{for } n \text{ a positive integer greater than 1 or equal to 1}$$

then X is complete.

$$E[h(X)] = \int_0^\infty \frac{h(x)}{\Gamma n\, 2^n} e^{-x/2} x^{n-1}\,dx = 0$$

implies that $h(x) x^{n-1}/\Gamma n\, 2^n = 0$ for every x by the unicity property of the Laplace transform. Hence $h(X) = 0$, a.e. Lebesgue, and so X is complete.

NOTE

Example 6 is a particular case of this density when $n = 1$. In that case X is complete for every positive real number λ.

EXAMPLE 9

Let X be a random variable with density function

$$f(x) = \frac{1}{g(T,\theta)\pi[1+(x/\theta)^2]\theta} \qquad \begin{array}{c} -T < x < T \\ -\infty < \theta < \infty \end{array}$$

where

$$g(T,\theta) = \int_{-T}^{T} \frac{dx}{\pi[1+(x/\theta)^2]\theta}$$

Then X is not complete.

If $h(X)$ is any odd function, then $E[h(X)] = 0$ and this does not imply that $h(X) = 0$, a.e. Lebesgue.

We shall now give general results concerning the completeness of families of probability measures (see Lehmann [4]).

*Theorem 2

Let X be a random vector with probability distribution

$$dp_\theta^x = k(\theta)\exp\left[\sum_{i=1}^{r} \theta_i T_i(x)\right] dp(x) \qquad (20)$$

and let p^t be the family of distributions of $t = [T_1(x), \ldots, T_r(x)]$, $\theta = (\theta_1, \ldots, \theta_r)$ as θ ranges over the set ω contained in euclidean r-space. Then p^t is complete provided ω contains an r-dimensional rectangle.

Proof:

Without loss of generality we assume that ω contains the rectangle

$$l = \{(\theta_1, \ldots, \theta_r): -c \leq \theta_i \leq c, i = 1, \ldots, r\}$$

Let $h(t) = h^+(t) - h^-(t)$ be such that

$$E_\theta[h(t)] = 0 \qquad \text{for all } \theta \in \omega$$

where $h^+(t)$, $h^-(t)$ are nonnegative functions. Then for all $\theta \in l$, if μ denotes the measure induced in t-space by the measure p,

$$\int e^{\Sigma \theta_i t_i} h^+(t) \, d\mu(t) = \int e^{\Sigma \theta_i t_i} h^-(t) \, d\mu(t) \qquad (21)$$

and hence for $\theta = 0$

$$\int h^+(t) \, d\mu(t) = \int h^-(t) \, d\mu(t)$$

Dividing h by a constant, one can take the common value of these two

integrals to be 1, so that
$$dp^+(t) = h^+(t)\,d\mu(t) \quad \text{and} \quad dp^-(t) = h^-(t)\,d\mu(t)$$
are probability measures, and
$$\int e^{\Sigma \theta_i t_i}\,dp^+(t) = \int e^{\Sigma \theta_i t_i}\,dp^-(t) \tag{22}$$
for all θ in l. We now assume that (22) remains valid when the θ_j are complex variables, that is, $\theta_j = u_j + iv_j$, $j = 1, \ldots, r$. For any fixed $\theta_1, \theta_2, \ldots, \theta_{j-1}, \theta_{j+1}, \ldots, \theta_r$ with real parts strictly between $-c$ and $+c$, then by the result of (Prob. 12) the integrals are analytic functions of θ_j in
$$R_j = \{\theta_j|\ -c < u_j < c,\ -\infty < v_j < \infty\}$$
of the complex plane. For $\theta_2, \ldots, \theta_r$ fixed, real, and between $-c$ and c, equality of the integrals holds on the line segment
$$\{(u_1,v_1),\ -c < u_1 < c,\ v_1 = 0\}$$
and can therefore be extended to R_1, in which the integrals are analytic. By induction the result can be extended to the region $\{(\theta_1, \ldots, \theta_r),\ \theta_j \in R_j \text{ for } j = 1, \ldots, r\}$. Hence, setting $u_j = 0, j = 1, \ldots, r$,
$$\int e^{i\Sigma v_j t_j}\,dp^+(t) = \int e^{i\Sigma v_j t_j}\,dp^-(t)$$
These are the characteristic functions of the distributions p^+ and p^-, respectively. Hence, by the uniqueness theorem for characteristic functions, the distributions p^+ and p^- coincide. Thus it follows that
$$h^+(t) = h^-(t) \quad \text{a.e.}$$
and so
$$h(t) = 0 \quad \text{a.e.}$$

REMARK

If one has completeness for a class of measures, one can in some cases infer completeness for a larger class. This fact is stated more precisely in the following theorem.

Theorem 3

Let $p^t = \{p_\theta{}^t,\ \theta \in \omega\}$ be a complete family of distributions, and let $\omega \subset \Omega$. Then $p^t = \{p_\theta{}^t,\ \theta \in \Omega\}$ is also complete provided none of the added measures assign positive probability to sets having zero probability for each measure of ω.

Proof:
Since the result is obvious, the proof is left to the reader.

Lemma 1 (*Halmos*)

If $S(p_1, \ldots, p_n)$ is a homogeneous polynomial of positive degree satisfying $S(p_1, \ldots, p_n) = 0$ whenever $0 \le p_i \le 1$, $i = 1, \ldots, n$, and $\sum_{1}^{n} p_i = 1$, then $S(p_1, \ldots, p_n)$ is zero identically.

Proof:
When $n = 1$, the lemma is obvious. The proof for arbitrary n is by induction, i.e., by demonstrating that if the result holds for $n - 1$, then it holds for n.

If we replace each p_i by dp_i, then from homogeneity a power of d will factor out, leaving the original polynomial. Therefore the restriction $0 \le p_i \le 1$ and $\Sigma p_i = 1$ may be replaced by the restriction $p_i \ge 0$, $i = 1, 2, \ldots, n$.

If we write $S(p_1, \ldots, p_n)$ as a polynomial in p_n, we have, for given $p_1, \ldots, p_{n-1}$, that it is identically zero for all $p_n \ge 0$. Hence the coefficients of the different powers of p_n must be zero. But since these coefficients are homogeneous functions of $n - 1$ variables, the lemma follows by induction.

Definition 3

Let us define $p(x)$ a distribution which assigns probability $p_1, p_2, \ldots, p_k$, respectively, to the disjoint intervals $l_1, l_2, \ldots, l_k$ on the real line with $\sum_{i=1}^{k} p_i = 1$; within each interval we assume that the distribution is uniform; i.e., we assume that the distribution has a density function which is constant-valued within each interval. We call such a distribution *uniform within intervals*.

**Theorem 4* (*Fraser*)

The order statistic $O(x_1, \ldots, x_k) = O(\mathbf{x})$ is complete for the class of distributions over R corresponding to each coordinate having the same distribution function $F(x)$, a distribution uniform within intervals.

Proof:

We have to show that any real function $f[O(\mathbf{x})]$ of the order statistic satisfying
$$E\{f[O(\mathbf{x})]\} = 0 \tag{23}$$
for all the given distributions is necessarily zero almost everywhere in Lebesgue measure. We first find a convenient way of expressing a function of the order statistic $O(\mathbf{x})$. It can be seen that any function of $O(\mathbf{x})$ is a symmetric function of the $x_1, x_2, \ldots, x_k$. Conversely, any symmetric function of the x_i's is independent of the particular order of the values and so is a function only of the order statistic $O(\mathbf{x})$. We therefore consider any symmetric function $f(x_1, \ldots, x_k)$ whose expectation is zero and prove that it is zero almost everywhere. We have

$$\begin{aligned} 0 &= E[f(x_1, \ldots, x_k)] \\ &= \sum_{i_1=1}^{k} \cdots \sum_{i_n=1}^{k} p_{i_1} \cdots p_{i_n} J(i_1, \ldots, i_k) \end{aligned} \tag{24}$$

where

$$J(i_1, \ldots, i_k) = \left[\prod_{r=1}^{k} L(i_r)\right]^{-1} \int_{l_{i_1}} \cdots \int_{l_{i_k}} f(x_{i_1}, \ldots, x_{i_k}) \prod_{i=1}^{k} dx_i \tag{25}$$

and $L(i_r)$ is the length of the interval l_{i_r}. Since $f(x_1, \ldots, x_k)$ is symmetric, so also is $J(i_1, \ldots, i_k)$. Therefore (24) can be written as

$$0 = \sum \prod_{r=1}^{k} p_r^{a_r} C(a_1, \ldots, a_k) \tag{26}$$

where the summation is over all nonnegative integers $a_1, \ldots, a_k$ summing to k, and where $C(a_1, \ldots, a_k)$ is an integral multiple of the $J(i_1, \ldots, i_k)$ with a_r of the arguments, i_α equal to r for $r = 1, 2, \ldots, k$.

The expression on the right-hand side of (26) satisfies the conditions of Lemma 1. Therefore

$$C(a_1, \ldots, a_k) = 0$$

and thus $J(i_1, \ldots, i_k) = 0$.

It follows then that

$$\int_{l_{i_1}} \cdots \int_{l_{i_k}} f(x_1, \ldots, x_k) \prod_{i=1}^{k} dx_i = 0 \tag{27}$$

for $i_r = 1, \ldots, k, r = 1, \ldots, k$, and all disjoint intervals $l_1, \ldots, l_k$.

68 PARAMETRIC ESTIMATION

Expression (27) determines a measure for all product sets $l_{i_1}, \ldots, l_{i_k}$, and this measure is zero.

Using the left- and right-hand sides of (27), respectively, we determine two measures for any Borel set A, which are

$$\int \cdots \int_A f(x_1, \ldots, x_k) \prod_{i=1}^{k} dx_i$$

and

$$\int \cdots \int_A O\Pi \, dx_i$$

But, since the two extensions of the measure must be identical, we have by the Radon-Nikodym theorem that $f(x_1, \ldots, x_k) = 0$ almost everywhere.

In the next section we consider a weaker property of completeness.

3 □ BOUNDEDLY COMPLETE FAMILIES OF PROBABILITY MEASURES

In many estimation problems one is interested in bounded estimates only, and so one is led to investigate the existence of bounded unbiased estimates of zero. We formally define the families of probability measures possessing only such trivial estimates.

Definition

The family p^x of measures is boundedly complete if

$$\int_R f(x) \, dp_\theta^x = 0 \qquad \text{for all } \theta \in \omega \tag{1}$$

and $f(x)$ bounded and real implies that $f(x) = 0$, a.e. p^x.

We illustrate this definition by an example and show that a family of probability measures can be boundedly complete without being complete.

EXAMPLE

Let ω be the open interval $0 < \theta < 1$. p_θ^x assigns the measure $(1 - \theta)^2 \theta^x$ to the points $x = 0, 1, \ldots$, measure θ to the point $x = -1$, and measure zero to the complement of this set of points. Condition

(1) then becomes

$$f(-1)\theta + \sum_{x=0}^{\infty} f(x)(1-\theta)^2 \theta^x = 0$$

Therefore,

$$\sum_{x=0}^{\infty} f(x)\theta^x = -f(-1)\theta(1-\theta)^{-2} = -f(-1) \sum_{x=0}^{\infty} x\theta^x \quad (2)$$

and so

$$\sum_{x=0}^{\infty} [f(x) + f(-1)x]\theta^x = 0 \quad (3)$$

Equation (3) implies that $f(x) + f(-1)x = 0$ for $x = -1, 0, 1, \ldots,$ and so either $f(-1)$ is not zero and the $f(x)$ is unbounded or $f(x)$ is zero for $x = -1, 0, 1, \ldots$. Hence, if $f(x)$ is bounded and satisfies (1), then $f(x) = 0$, a.e. p^x, and thus p^x is boundedly complete. On the other hand, if $f(x) = -f(-1)x$ and $f(-1) \neq 0$, then (1) is satisfied but $f(x) \neq 0$, a.e. p^x, and so p^x is not complete.

Theorem

If a family p^x of measures is complete, then it is boundedly complete.

Proof:

The result follows immediately from the definitions.

4 □ STRONGLY COMPLETE FAMILIES OF PROBABILITY MEASURES

We next consider a stronger property than completeness, one which implies completeness. We define the property.

Definition

The family of probability measures $p^T = \{p_\theta^T | \theta \in \omega\}$ is strongly complete with respect to a measure m on ω if for any subset ω^* of ω for which $m(\omega - \omega^*) = 0$ the condition

$$E_\theta[f(T)] = 0 \quad (4)$$

for all $\theta \in \omega^*$ and any real statistic $f(t)$ implies that $f(t) = 0$ a.e. p_ω^T.

Thus strong completeness requires that any unbiased estimate of zero for a subfamily of probability measures is zero almost everywhere with respect to the full family, provided the measures omitted form a set having m measure zero. This concept leads to the following theorem concerning families M^x of product measures $m_\theta{}^x$ (not necessarily probability measures) where the whole space is W^x.

*Theorem   (Lehmann and Scheffé)

Let $x = (x',x'')$, $W^x = W^{x'} \times W^{x''}$, $F^x = F^{x'} \times F^{x''}$, $\theta = (\theta',\theta'')$, $\omega = \omega' \times \omega''$, $\mathcal{L}^\theta = \mathcal{L}^{\theta'} \times \mathcal{L}^{\theta''}$, $L^\theta = L^{\theta'} \times L^{\theta''}$.

Suppose $M^x = \{m_\theta{}^x, \theta \in \omega\}$, where $dm_\theta{}^x = g_\theta(x)\, d\mu^x$, μ^x is σ-finite,

$$g_\theta(x) = g'_{\theta'}(x') g''_{\theta''}(x'')$$

where $g'_{\theta'}(x')$ is measurable ($\mathcal{L}^{\theta'} \times F^{x'}$) and $g''_{\theta''}(x'')$ is measurable ($\mathcal{L}^{\theta''} \times F^{x''}$). Suppose further that there exists a measure $\mu^{x''}$ on $F^{x''}$ and conditional measure $\mu^{x'|x''}$ on $F^{x'}$ depending on x'' such that whenever the integral

$$\int_{W^x} f(x',x'') g_\theta(x)\, d\mu^x \tag{5}$$

is defined, it equals the iterated integral

$$\int_{W^x} F(\theta',x'') g''_{\theta''}(x'')\, d\mu^{x''} \tag{6}$$

where

$$F(\theta',x'') = \int_{W^{x'}} f(x',x'') g'_{\theta'}(x')\, d\mu^{x'|x''} \tag{7}$$

is measurable ($\mathcal{L}^{\theta'} \times F^{x''}$). Then M^x is strongly complete providing the same is true of $M^{x''} = \{m'_{\theta'',}\theta'' \in \omega''\}$ and (a.e. $\mu^{x''}$) of

$$M^{x'|x''} = \{m_{\theta}{}^{x'|x''} | \theta' \in \omega'\},$$

where $dm_{\theta'',}{}^{x''} = g''_{\theta''}(x'')\, d\mu^{x''}$ and $dm_{\theta',}{}^{x'|x''} = g'_{\theta'}(x')\, d\mu^{x'|x''}$.

Proof:

We have to show that if the integral (5) vanishes a.e. L^θ, then

$$f(x',x'') = 0,$$

a.e. M^x. Let N be the set in ω on which the integral (5) does not vanish, so that $N \in \mathcal{L}^\theta$ and $L^\theta(N) = 0$. Then by the hypothesis of the theorem the iterated integral (6) exists and equals zero for $\theta \notin N$. Let S be the set of points (θ',x'') such that where $F(\theta',x'')$ defined in (7) does not vanish for $S \in \mathcal{L}^{\theta'} \times F^{x''}$. We show first that the product measure $L^{\theta'} \times M_{\theta'',}{}^{x''}$

of S is zero for all $\theta'' \in \omega''$. Denote by $N_{\theta'}$ the section of N by $\theta' = $ const, that is, $N_{\theta'} = \{\theta''|(\theta',\theta'') \in N\}$. Since $(L^{\theta'} \times L^{\theta''})(N) = 0$, $L^{\theta''}(N_{\theta'}) = 0$ except for $\theta' \in N'$, where $N' \in \mathcal{L}^{\theta'}$ and $L^{\theta'}(N') = 0$. For $\theta' \notin N'$ we now have

$$\int_{W_{x''}} F(\theta',x'')\, dm_{\theta'',x''} = 0 \qquad \text{a.e. } L^{\theta''}$$

and so by the strong completeness of $M^{x''}$, $F(\theta',x'') = 0$, a.e. $M^{x''}$. Thus all sections of S by $\theta' = $ const with $\theta' \notin N'$ have $m_{\theta'',x''}$ measure zero for all $\theta'' \in \omega''$. Hence $(L^{\theta'} \times m''_{\theta''})(S) = 0$ for all $\theta'' \in \omega''$.

We may now conclude that if $S_{x''}$ denotes a section of S by $x'' = $ const, then the set N'' in $W^{x''}$ for which $L^{\theta'}(S_{x''}) \neq 0$ has $m_{\theta'',x''}$ measure zero for all $\theta'' \in \omega''$. In other words, except for $x'' \in N''$, a null set of $M^{x''}$,

$$F(\theta',x'') = \int_{W_{x'}} f(x',x'')\, dm_{\theta',x'|x''} = 0 \qquad \text{a.e. } L^{\theta'}$$

Hence by strong completeness of $M^{x'|x''}$, we have for $x'' \in N''$, $f(x',x'') = 0$, a.e. $M^{x'|x''}$, and from this we now conclude that $f(x',x'') = 0$, a.e. m^x.

For further discussion of the subject, refer to [2] and [4] to [6].

PROBLEMS

1. Let $X_1, X_2, \ldots, X_n$ be independent random variables, each uniformly distributed on $[0,\theta]$, where $\theta \in \omega$, the set of positive real numbers. Prove that $t(X_1, \ldots, X_n) = \max_{1 \leq i \leq n} X_i$ is complete.

2. Let X be hypergeometric (n,D,N), where n and N are fixed positive integers, $D \in \{0,1,2,\ldots,N\}$. Prove that $t(X) = X$ is complete.

3. Let X be hypergeometric (n,D,N), where n and D are fixed, $N \in \omega$, the set of positive integers greater than max (n,D). Is X complete?

4. Let $X_1, X_2, \ldots, X_n$ be independent random variables normally distributed with unknown mean μ and unknown variance σ^2. Construct complete statistics for the parametric upper half plane (μ,σ^2). *Hint:* Use Theorem 2 of Sec. 2.

5. Let $X_1, X_2, \ldots, X_n$ be independent random variables normally distributed with mean 0 and unknown variance $\sigma^2 \in \omega$, the set of positive real numbers. Prove that $t(X_1, \ldots, X_n) = \sum_1^n X_i^2$ is complete.

72 PARAMETRIC ESTIMATION

6. Let X_1, X_2 be independent random variables where X_i is binomial (n, p_i) for $i = 1, 2$, and $(p_1, p_2) = \omega$, the interior of the unit square. Prove that $t(X_1, X_2) = (X_1, X_2)$ is complete.

7. Let
$$X_{11}, X_{12}, \ldots, X_{1n_i}$$
$$X_{21}, X_{22}, \ldots, X_{2n_i}$$
be independent where X_{ij} is normal (μ_i, σ_i^2), $j = 1, 2, \ldots, n$, $i = 1, 2$, $(\mu_1, \mu_2, \sigma_1^2, \sigma_2^2) \in \omega$, where $\omega = \{(\mu_1, \mu_2, \sigma_1^2, \sigma_2^2), -\infty < \mu_i < \infty \text{ for } i = 1, 2\}$. Prove that
$$t(X_{11}, \ldots, X_{2n_2}) = \left(\sum_{j=1}^{n_1} X_{i_j}, \sum_{j=1}^{n_2} X_{2_j}, \sum_{1}^{n_1} X_{1_j}^2, \sum_{j=1}^{n_2} X_{2_j}^2 \right)$$
is complete.

8. Consider again the setup of Prob. 7 but with $\sigma_1^2 = \sigma_2^2$. Prove that
$$t(X_{11}, \ldots, X_{2n}) = \left(\sum_{1}^{n_1} X_{1_j}, \sum_{j=1}^{n_2} X_{2_j}, \sum_{i=1}^{2} \sum_{j=1}^{n_i} X_{i_j}^2 \right)$$
is complete.

9. Let $X_1, X_2, \ldots, X_n$ be independent random variables with common density function
$$g(x; \theta, \eta) = \begin{cases} \dfrac{1}{\eta - \theta} & \theta \leq x \leq \eta \\ 0 & \text{elsewhere} \end{cases}$$
where θ and η are real numbers such that $\theta < \eta$. Then consider a statistic
$$t(X_1, \ldots, X_n) = \max_{1 \leq i \leq n} X_i - \min_{1 \leq i \leq n} X_i$$
Is t complete or boundedly complete?

10. (a) (Lehmann and Scheffé) Suppose $M^{x'} = \{m_{\theta'}^{x'}; \theta' \in \omega'\}$ and $M^{x''} = \{m_{\theta''}^{x''}, \theta'' \in \omega''\}$, where $dm_{\theta'}^{x'} = g'_{\theta'}(x') d\mu^{x'}$ and
$$dm_{\theta''}^{x''} = g''_{\theta''}(x'') d\mu^{x''},$$
are strongly complete families of measures. Let $x = (x', x'')$, $\omega^x = \omega^{x'} \times \omega^{x''}$, $F^x = F^{x'} \times F^{x''}$, $\theta = (\theta', \theta'')$, $\omega = \omega' \times \omega''$, $\mathcal{L}^\theta = \mathcal{L}^{\theta'} \times \mathcal{L}^{\theta''}$, $L^\theta = L^{\theta'} \times L^{\theta''}$, and let $M^x = \{m_\theta^x, \theta \in \omega\}$, where $m_\theta^x = m_{\theta'}^{x'} \times m_{\theta''}^{x''}$. Prove that the family M^x of product measures is strongly complete.

(b) If $M^x = \{m_\theta^x; \theta \in \omega\}$ is strongly complete, and if $R^x = \{k_\theta^x | \theta \in \omega\}$ is defined by $dk_\theta^x = u(\theta)u(x) \, dM_\theta^x$, where $u(\theta)$ and $u(x)$ are nonnegative, $u(\theta)$ is measurable $(\mathcal{L}^\theta)$, $u(\theta) \neq 0$, a.e. L^θ, and $u(x)$ is measurable (F^x), prove that R^x is strongly complete.

11. Let $p_\omega^t = \{p_\theta^t | \theta \in \omega\}$ be a complete family of probability measures induced by a statistic t over the Borel field F, and let $p_\Omega^{t'} = \{p^{t'}, \xi \in \Omega\}$ be a strongly complete family of probability measures induced by a statistic t' over the Borel field F'; then prove that

$$\int_{F \times F'} f(t,t') \, dp_\theta^T(t) \, dp_\xi^{T'}(t) = 0 \quad \text{for all } (\theta, \xi)$$

implies that $f(t,t') = 0$, a.e. $\{p_\theta^T \times p_\xi^{T'} | (\sigma, \xi) \in \omega \times \Omega\}$. For the proof, see Fraser [2].

12. Let h be any bounded measurable function on a probability space. Then

$$\int h(x) \exp\left(\sum_{k=1}^r \theta_k T_k^{(x)}\right) dp(x) \tag{1}$$

which is a function of the complex variables

$$\theta_k = u_k + iv_k \qquad k = 1, 2, \ldots, r$$

Show that it is an analytic function in each of the variables in the parametric euclidean space, and prove that the derivation of all orders with respect to the θ's of (1) can be computed under the integral sign. For the proof see Lehmann [4, p. 52].

REFERENCES

1 FRASER, D. A. S.: Completeness of the Order Statistics, *Can. J. Math.*, vol. 6, pp. 42–45, 1953.
2 FRASER, D. A. S.: "Nonparametric Methods in Statistics," John Wiley & Sons, Inc., New York, 1957.
3 HALMOS, P. R.: "Measure Theory," D. Van Nostrand Company, Inc., Princeton, N.J., 1950.
4 LEHMANN, E. L.: "Testing Statistical Hypotheses," John Wiley & Sons, Inc., New York, 1959.
5 LEHMANN, E. L., and H. SCHEFFÉ: Completeness, Similar Regions and Unbiased Estimation, *Sankhyā*, ser. A, vol. 10, pp. 305–340, 1950.
6 LEHMANN, E. L., and H. SCHEFFÉ: Completeness, Similar Regions and Unbiased Estimation II, *Sankyā*, ser. A, vol. 15, pp. 219–236, 1955.

chapter Five

Sufficiency

1 □ INTRODUCTION

The concept of sufficiency is of paramount importance in mathematical and applied statistics. When an experimenter collects observations of his experiment, eventually he is confronted with the task of constructing a function of his observations to obtain some desired information. Thus Fisher introduced the concept of a sufficient statistic as one "containing all the relevant information." Neyman gave a method of finding sufficient statistics. We define sufficient statistics and give the Neyman criterion in Sec. 2. Halmos and Savage [11] extended Neyman's result in the dominated case, the subject of Sec. 3. Lehmann and Scheffé [13] considered minimal sufficient statistics and many other aspects of sufficiency. Then in a series of papers Bahadur [1–3] investigated the concept deeply and in an interesting manner in terms of subfields and posed many fascinating problems. Limitations of space, unfortunately, prevent discussion of his results here. Burkholder [6, 7], in addition to answering several problems presented by Bahadur, extended the concept of sufficiency from the dominated to the undominated case. Problems concerning minimal dimensions have been studied by Barankin and Katz [4], while Fraser [8], drawing on contributions of Koopman [12] and others, introduced the concept of local conditional sufficiency.

2 □ SUFFICIENT STATISTICS

Definition

A statistic t is said to be sufficient for the family $P = \{p_\theta^x, \theta \in \omega\}$, or simply sufficient for θ, where t is a Borel function of X and conditional distribution $P_\theta^{x|t}$ exists almost surely for each θ, if this conditional distribution of X given $t = y$ is independent of θ.

EXAMPLE 1

Let $X_1, X_2, \ldots X_n$ be independent random variables, each having density function

$$p(X_i = 1) = p \quad\quad 0 < p < 1$$
$$\quad\quad\quad\quad\quad\quad\quad i = 1, \ldots, n$$
$$p(X_i = 0) = 1 - p$$

Then the total number of successes, $\sum_{i=1}^{n} X_i$, is a sufficient statistic for p. For if $t = y$, exactly $\binom{n}{y}$ outcomes $(x_1, \ldots, x_n)$ of the experiment are possible and each of these has probability

$$\frac{p^y q^{n-y}}{\binom{n}{y} p^y q^{n-y}} = \frac{1}{\binom{n}{y}}$$

which is independent of p. Hence t is sufficient for p.

EXAMPLE 2

Let $X_1, X_2, \ldots, X_n$ be independent random variables with density function

$$p_N(X_i = k) = \frac{1}{N} \quad k = 1, 2, \ldots, N$$

$$N \in \omega$$

the set of positive integers. Then $t = \max_{1 \leq i \leq n} X_i$ is a sufficient statistic. One can easily verify that the density function of t is

$$p_N[t(X) = y] = p_N[t(X) \leq y] - p_N[t(X) < y] \quad y = 1, 2, \ldots, n$$

$$= \frac{y^n}{N^n} - \frac{(y-1)^n}{N^n}$$

$$p_N\{X_1 = x_1, \ldots, X_n = x_n | t(x) = y\}$$

$$= \frac{p_N\{X_{12} = x_1, \ldots, X_N = x_n; t(x) = y\}}{p_N[t(x) = y]}$$

$$= \frac{p_N(X_1 = x_1, \ldots, X_n = x_n)}{p_N[t(x) = y]}$$

$$= \frac{1/N^n}{y^n/N^n - (y-1)^n/N^n}$$

$$= \frac{1}{y^n - (y-1)^n}$$

Hence t is sufficient for N.

EXAMPLE 3

Let $X_1, X_2, \ldots, X_n$ be independent random variables with density function

$$p(X_i = k_i) = e^{-\lambda} \frac{\lambda^{k_i}}{k_i!} \quad \text{for } i = 1, \ldots, n$$

$$\lambda > 0$$

k a nonnegative integer

Then $t = \sum_1^n X_i$ is a sufficient statistic for λ. We know that the density function of t is

$$p(t = y) = \frac{e^{-n\lambda}(n\lambda)^y}{y!}$$

where y is a nonnegative integer.

$$\{X_1 = k_1, \ldots, X_n = k_n | t = y\} = \frac{p(X_1 = k_1, \ldots, X_n = k_n, t = y)}{p(t = y)}$$

$$= \frac{p(X_1 = k_1, \ldots, X_n = k_n)}{p(t = y)}$$

$$= \frac{e^{-n\lambda} \lambda^T y!}{\prod_{i=1}^{n} k_i! e^{-n\lambda} (n\lambda)^y} \quad \text{where } T = \sum_{i}^{n} k_i$$

$$= \frac{y!}{\prod_{i=1}^{n} k_i! n^y}$$

Hence t is sufficient for λ.

Now we consider a method by which one can check whether a statistic is sufficient or not. In order to fix the idea we discuss it for discrete density functions.

Factorization criterion (Neyman [14])

Let $p_\theta(x) = p_\theta(X = x)$. Then a necessary and sufficient condition for t to be sufficient for θ is that there exist a factorization

$$p_\theta(x) = g_\theta[t(x)]\phi(x) \tag{1}$$

where the first factor may depend on θ but depends on x only through $t(x)$ while the second factor is independent of θ such that for

$$A = \{x : t(x) = y\},$$

$\sum_{x \in A} \phi(x) > 0$, $p_\theta(t = y) > 0$ for every value y of t and $\phi(x) \geq 0$.

Proof:

Suppose that (1) holds, and let $t(x) = y$. Then $p_\theta(t = y) = \Sigma p_\theta(x')$ summed over all x' with $t(x') = y$ and the conditional probability

$$p_\theta(X = x | t = y) = \frac{p_\theta(x)}{p_\theta(t = y)} = \frac{g_\theta[t(x)]\phi(x)}{\Sigma g_\theta(t)\phi(x')} = \frac{\phi(x)}{\Sigma \phi(x')}$$

which is independent of θ. Conversely, if this conditional distribution

78 PARAMETRIC ESTIMATION

does not depend on θ and is equal to, say, $h(x,y)$, then

$$p_\theta(x) = p_\theta(t = y) h(x,y)$$

and so (1) holds. For the continuous case see Prob. 12.

EXAMPLE 4

Let $X_1, \ldots, X_n$ be independent, each being $R(0,\vartheta)$. The probability density of the sample is

$$p_\theta(x_1, \ldots, x_n) = \begin{cases} \dfrac{1}{\theta^n} & \text{if all } x_i \in [0,\theta] \\ 0 & \text{otherwise} \end{cases}$$

Let us define a function f as

$$f(a,b) = \begin{cases} 1 & \text{if } b \geq a \\ 0 & \text{if } a > b \end{cases}$$

Let
$$y = \max(X_1, \ldots, X_n) \qquad z = \min(X_1, \ldots, X_n)$$

Then
$$p_\theta(x_1, \ldots, x_n) = \frac{f(y,\theta)}{\theta^n} f(0,z)$$

It follows from the Neyman criterion that $y = \max(X_1, \ldots, X_n)$ is sufficient for θ.

On the other hand, using the definition directly,

$$p_\theta\{x_1, \ldots, x_n | y = t\} = \frac{1}{\theta^n} \frac{\theta^n}{n t^{n-1}} = \frac{1}{n t^{n-1}}$$

and so y is sufficient for θ.

3 □ FAMILIES OF DISTRIBUTIONS ADMITTING SUFFICIENT STATISTICS

As Fraser [10] notes, the outcome of an experiment forms a sufficient statistic but one which we hope to improve by reduction to a statistic of smaller dimension. For several well-known models the reduction can be quite extreme; for a set of observations, however many, on a normal distribution of unknown mean and variance, the sum of the observations, the sum of their squares, and the total number of observations form a sufficient statistic; for a Poisson distribution, the sum of the observations and the number of observations are sufficient. Unfortunately, distributions for which the dimension of some sufficient statistic is independent of the number of observations belong in general to a special family of distributions. The nature of this family was first investigated by Koopman [12], who considered sufficient statistics of dimension 1. Barankin and Maitra [5] have studied distributions with minimal sufficient statistics of higher dimension and have shown that the distributions are of exponential form again. We consider the theorem of Koopman.

Theorem

Let $X_1, \ldots, X_n$ be independent random variables with the common density function $f_\theta(x)$ such that

$$\prod_{i=1}^{n} f_\theta(x_i) = g_\theta(t) h(x_1, \ldots, x_n)$$

and f, g are differentiable with respect to θ and the x_i's and the range of the x_i's does not depend on θ.
Then

$$\prod_{i=1}^{n} f_\theta(x_i) = h(x_1, \ldots, x_n) e^{\phi_1(\theta) k(t) + \phi_2(\theta)}$$

Proof:

$$\prod_{i=1}^{n} f_\theta(x_i) = g_\theta(t) h(x_1, \ldots, x_n)$$

Therefore

$$\sum_{i=1}^{n} \frac{\partial \log f_\theta(x_i)}{\partial \theta} = \frac{\partial \log g_\theta(t)}{\partial \theta} = k_\theta(t) \qquad (2)$$

Since (2) is true for all θ, we choose an arbitrary θ_0 to obtain

$$r = \sum_{i=1}^{n} v(x_i) = k_{\theta_0}(t) = k(t)$$

where

$$v(x_i) = \frac{\partial \log f_{\theta_0}(x_i)}{\partial \theta_0}$$

Assuming that $k(t)$ satisfies the necessary conditions of the inverse function theorem, we obtain
$$t = k^{-1}(r) = J(r)$$
and define $q_\theta(r) = k_\theta[k^{-1}(r)]$. Differentiating (2) with respect to x_i, for i arbitrary, we obtain
$$\frac{\partial^2}{\partial x_i \, \partial \theta} \log f_\theta(x_i) = \frac{\partial q_\theta(r)}{\partial r} \frac{\partial r}{\partial x_i}$$
Therefore
$$\frac{\partial^2 \log f_\theta(x_i)}{\partial x_i \, \partial \theta \; \partial r / \partial x_i} = \frac{\partial q_\theta(r)}{\partial r} \tag{3}$$

Now on the left-hand side of (3), the numerator is a function of x_i and θ while the denominator is a function of x_i only. The right-hand side is a function of θ and of r only, and r is a symmetric function of $x_1, \ldots, x_n$. Thus (3) must be a function of θ alone, say
$$\frac{\partial q_\theta(r)}{\partial r} = C_1(\theta)$$
and so
$$q_\theta(r) = rC_1(\theta) + C_2(\theta)$$
or
$$k_\theta(t) = k_{\theta_0}(t)C_1(\theta) + C_2(\theta)$$
$$= k(t)C_1(\theta) + C_2(\theta)$$
$$= \frac{\partial \log g_\theta(t)}{\partial \theta}$$
Therefore
$$g_\theta(t) = e^{\phi_1(\theta)k(t)+\phi_2(\theta)}$$
and so
$$\prod_{i=1}^{n} f_\theta(x_i) = h(x_1, \ldots, x_n) e^{\phi_1(\theta)k(t)+\phi_2(\theta)}$$

the desired result.

More insight into the implications of sufficiency can be obtained by considering the left-hand side of (3) as follows:
$$\frac{\partial^2}{\partial x_i \, \partial \theta} \log f_\theta(x_i) = C_1(\theta) \frac{\partial r}{\partial x_i}$$
$$= C_1(\theta) \frac{dv(x_i)}{dx_i}$$

and so
$$\frac{\partial}{\partial \theta} \log f_\theta(x_i) = C_1(\theta) v(x_i) + C_2(\theta)$$

Therefore
$$\log f_\theta(x_i) = \phi_1(\theta) v(x_i) + \phi_2(\theta) + w(x_i)$$

and so
$$f_\theta(x_i) = h(x_i) e^{\phi_1(\theta) v(x_i) + \phi_2(\theta)}$$

$$\prod f_\theta(x_i) = \prod h(x_i) \exp \phi_1(\theta) \sum_{i=1}^{n} v(x_i) + \phi_2(\theta)$$

Note that $\sum_{i=1}^{n} v(x_i)$ is a sufficient statistic.

4 □ SUFFICIENCY IN THE DOMINATED CASE

Definition *(Halmos and Savage [11])*

t is sufficient for the family of probability measures $\{p_\theta | \theta \in \omega\}$, defined on F, the family of Borel sets in the sample space X, provided that if f is a function, p_θ-integrable for each $\theta \in \omega$, then there is a function g, $(p_\theta t^{-1})$-integrable for each $\theta \in \omega$, such that

$$\int_{t^{-1}(B)} f(x)\, dp_\theta(x) = \int_B g(y)\, dp_\theta t^{-1}(y)$$

for all $B \in \beta$, the family of Borel sets in the space of values of t, the images under t of the Borel sets of X, and all $\theta \in \omega$.

$$g = E_\theta(f|t) \text{ is independent of } \theta.$$

In particular, taking $f(x)$ to be the indicator function for A, I_A, $A \in F$,

$$p_\theta[A \cap t^{-1}(B)] = \int_{t^{-1}B} I_A(x)\, dp_x$$
$$= \int_B g(y)\, dp_\theta\, t^{-1}(y) \qquad \text{for all } B \in \beta,\ \theta \in \omega$$

but $p_\theta[A \cap t^{-1}(B)] = \int_B p(A|y)\, dp_\theta\, t^{-1}(y)$, and so $g(y) = p(A|y)$, independent of θ, a.e. p.

Lemma 1

Suppose that g is a function from y into the real numbers. Then a necessary and sufficient condition that g is measurable with respect to β is that the function $g(t)$ is measurable with respect to F.

Proof:

Suppose g is measurable with respect to β. Then

$$\{y|g(y) \leq c\} \in \beta$$

implies that $g^{-1}\{(-\infty,c]\} \in \beta$.

$$[g(t)]^{-1}\{(-\infty,c]\} = t^{-1}(g^{-1}(-\infty,c]) \in F$$

Therefore, $g(t)$ is measurable with respect to F. Similarly we can see that $g(t)$ being measurable with respect to F implies that g is measurable with respect to β.

Lemma 2

If g is a (pt^{-1}) integrable function, then $g[t(x)]$ is p-integrable and

$$\int_{t^{-1}(B)} g[t(x)]\, dp(x) = \int_B g(y)\, dp\, t^{-1}(y) \qquad \text{for all } B \in \beta$$

Proof:

We assume sample spaces (X,F,p) and (T,β,pt^{-1}), the image under t of the initial sample space X.

Let us prove the result when g is the indicator function. Then the result will be true for simple functions and so true for measurable functions.

$$g(y) = \begin{cases} 1 & \text{if } y \in C \in \beta \\ 0 & \text{otherwise} \end{cases}$$

$$\begin{aligned}
\int_B g(y)\, dp\, t^{-1}(y) &= pt^{-1}(B \cap C) = p[t^{-1}(B \cap C)] \\
&= \int_{t^{-1}(B \cap C)} dp(x) \\
&= \int_{t^{-1}(B) \cap t^{-1}(C)} dp(x) \\
&= \int_{t^{-1}(B)} \text{indicator function of } t^{-1}(C)\, dp(x) \\
&= \int_{t^{-1}(B)} g[t(x)]\, dp(x)
\end{aligned}$$

and the proof is complete.

Halmos and Savage [11] proved the following result.

Let μ be a σ-finite measure on (X,F), and let $\{f_\theta | \theta \in \omega\}$ be a set of generalized probability density functions with respect to μ.

Let
$$p_\theta(A) = \int_A f_\theta(x) \, d\mu(x) \qquad A \in F$$

Then there exists a countable subset $(\theta_1, \theta_2, \ldots)$ of ω and positive real numbers, $c_1, c_2, \ldots$, such that if
$$\lambda = \sum_{i=1}^{\infty} c_i p_{\theta_i}$$

$\{p_\theta | \theta \in \omega\}$ is dominated by λ. That is,
$$\lambda(A) = 0 \Leftrightarrow p_\theta(A) = 0 \qquad \text{for all } \theta \in \omega$$

and there exist $\{f'_\theta | \theta \in \omega\}$ such that
$$p_\theta(A) = \int_A f'_\theta(x) \, d\lambda(x) \qquad A \in F$$

Therefore, without loss of generality, we shall assume that
$$\mu = \sum_{i=1}^{\infty} c_i p_{\theta_i} \tag{1}$$

Theorem 1 Factorization Theorem (Halmos and Savage [11])

A necessary and sufficient condition that a statistic t be sufficient for $\{p_\theta | \theta \in \omega\}$ is that there exist a nonnegative function h measurable with respect to F and μ-integrable, and a set $\{g_\theta | \theta \in \omega\}$ of nonnegative functions such that for each $\theta \in \omega$, $g_\theta(t)$ is measurable with respect to $t^{-1}F = F_0$ and $P\{x | h(x) > 0\} = 1$ and $h(x)$ is μ-integrable.

$$f_\theta(x) = g_\theta(t) h(x) \qquad \text{a.e. } \mu \text{ for all } \theta \in \omega \tag{2}$$

or
$$dp_\theta(x) = g_\theta[t(x)] d\mu(x)$$

Proof:

Let F_0 be the subfield of F induced by t, that is, the subfield of F such that elements of X with the same value of t occur only in common sets, and let t be sufficient for θ. Hence for every $\theta \in \omega$, $A \in F$, and $A_0 \in F_0$, we have

$$\int_{A_0} p\{A | t(x)\} \, dp_\theta(x) = p_\theta(A \cap A_0) \tag{3}$$

Therefore
$$\int_{A_0} p\{A|t(x)\} \, d\mu(x) = \mu(A \cap A_0)$$

because $\mu = \Sigma c_j p_\theta$ from (1). Let

$$g_\theta[t(x)] = \frac{dp_\theta(x)}{d\mu(x)} \quad \text{for } (F_0, \mu)$$

Note that g_0 is a function of $t(x)$, since it is defined for (F_0, μ).

We have to show that $g_\theta[t(x)]$ is the Radon-Nikodym derivative of p_θ for (F, μ) as well as for (F_0, μ). Let $A_0 = X$.

$$\begin{aligned}
p_\theta(A) &= \int p\{A|t(x)\} \, dp_\theta(x) \\
&= \int E_\mu\{I_A(y)|t(y) = t(x)\} \, dp_\theta(x) \\
&= \int E_\mu\{I_A(y)|t(y) = t(x)\} g_\theta[t(x)] \, d\mu(x) \\
&= \int E_\mu\{I_A(y) g_\theta[t(y)]|t(y) = t(x)\} \, d\mu(x) \\
&= E\{I_A(y) g_\theta[t(y)]\} \\
&= \int_A g_\theta[t(y)] \, d\mu(x)
\end{aligned}$$

Note that we used the fact that $E\{E(Y|X)\} = E(Y)$.

Let (2) hold. Then the conditional probability $p_\mu(A|t)$ is a conditional-probability function for all $\{p_\theta|\theta \in \omega\}$. Let $g_\theta[t(x)] = dp_\theta(x)/d\mu(x)$ on F_0, and for any fixed A and θ let $dm = I_A \, dp_\theta$. Then over F_0,

$$\frac{dm(x)}{dp_\theta(x)} = E_\theta[I_A(x)|t(x)]$$

and hence over F_0

$$\frac{dm(x)}{d\mu(x)} = p_\theta[A|t(x)] g_\theta[t(x)]$$

Therefore

$$p_\mu[A|t(x)] g_\theta[t(x)] = p_\theta[A|t(x)] g_\theta[t(x)]$$

for (F_0, μ) and hence for (F_0, p_θ). But $g_\theta[t(x)] \neq 0$, a.e. (F_0, p_θ). Hence $p_\theta[A|t(x)] = p_\mu[A|t(x)]$, (F_0, p_θ). Hence the desired result.

EXAMPLE 1

Let $X_1, X_2, \ldots, X_n$ be independent random variables, each one normally distributed with unknown mean $-\infty < \eta < \infty$ and known variance σ^2. We shall show that $\bar{X} = \frac{1}{n} \sum_{1}^{n} X_i$ is a sufficient statistic for η.

We have the density function

$$\prod_{i=1}^{n} f_\eta(x_i) = \frac{1}{\sigma^n(2\pi)^{n/2}} \exp\left[-\frac{\Sigma(x_i - \eta)^2}{2\sigma^2}\right]$$

$$= \exp\left[-\frac{1}{2\sigma^2}\Sigma(x_i - \bar{x})^2 - \frac{n\bar{x}^2}{2\sigma^2}\right]$$

$$\frac{1}{\sigma^n(2\pi)^{n/2}} \exp\left[-\frac{n}{2\sigma^2}(\bar{x} - \eta)^2 + \frac{n}{2\sigma^2}\bar{x}^2\right]$$

$$= h(x_1, \ldots, x_n) g_\eta(\bar{x})$$

One can easily check that $h(x_1, \ldots, x_n)$ is integrable. This satisfies the condition of the theorem; hence $\bar{X}$ is a sufficient statistic for η.

Theorem 2

If X_i has distribution $\{p_{\theta_i}{}^{x_i} | \theta_i \in \omega_i\}$ on (W_i, F_i), $i = 1, 2$, and if $t_i(x_i)$ is sufficient for $\theta_i \in \omega_i$, then $[t_1(x_1), t_2(x_2)]$ is sufficient for the class of product measures given by $(\theta_1, \theta_2) \in \omega_1 \times \omega_2$ (under the assumption that the conditional probabilities are measures).

Proof:

From the assumption that $[t_1(x_1), t_2(x_2)]$ are sufficient we have, for all A_1, A_2, B_1, B_2,

$$p_{\theta_1}{}^{x_1}[A_1 \cap t_1^{-1}(B_1)] = \int_{B_1} p_1(A_1|s) \, dp_{\theta_1}{}^{t_1^{-1}}(s)$$

$$p_{\theta_2}{}^{x_2}[A_2 \cap t_2^{-1}(B_2)] = \int_{B_2} p_2(A_2|s) \, dp_{\theta_2}{}^{t_2^{-1}}(s)$$

By assumption, $p_1(A_1|t_1)$, $p_2(A_2|t_2)$ are measures over W_1, W_2. Let $p[A|(t_1,t_2)]$ be the product measure over $x_1 \times x_2$. Then, if $W = W_1 \times W_2$, $t = (t_1, t_2)$, $\theta = (\theta_1, \theta_2)$, $x = (x_1, x_2)$, we have

$$p_\theta{}^x[A \cap t^{-1}(B)] = \int_B p(A|s) \, dp_\theta{}^{t^{-1}}(s) \tag{4}$$

for all $A = A_1 \times A_2$ and $B = B_1 \times B_2$.

But both sides of the above equation define, by means of (4), a product measure on W. The measures are identical for all product sets $A = A_1 \times A_2$; therefore they agree for any A in the product space σ-algebra. This implies that (4) holds for all $A \in (F_1, F_2)$ and for all $B = B_1 \times B_2$. A similar argument gives equality for all $B \in (\beta_1, \beta_2)$ and establishes that (t_1, t_2) is sufficient for $(\theta_1, \theta_2) \in \omega_1 \times \omega_2$.

EXAMPLE 2

Let $\mathbf{x} = (X_1, \ldots, X_n)$, where the X_i are independent and each is normally distributed with mean ξ and variance σ^2. Similarly let

$$\mathbf{y} = (Y_1, \ldots, Y_m),$$

where the Y_i are independent and each is normally distributed with mean η and variance τ^2.

$$\left(\bar{X}, \frac{\Sigma(X_i - \bar{X})^2}{n-1}\right)$$

is a sufficient statistic for (ξ, σ^2) and

$$\left(\bar{Y}, \frac{\Sigma(Y_i - \bar{Y})^2}{m-1}\right)$$

is a sufficient statistic for (η, τ). Then by the above theorem

$$\left[\bar{X}, \bar{Y}, \frac{\Sigma(X_i - \bar{X})^2}{n-1}, \frac{\Sigma(Y_i - \bar{Y})^2}{m-1}\right]$$

is a sufficient statistic for the combined experiment with $\xi, \eta \in [-\infty, \infty]$ and $\sigma^2, \tau^2 \in [0, \infty]$.

Sufficiency of order statistics[1]

Let X be a random variable from the probability space $\{R, A, p_\theta | \theta \in \omega\}$. If the experiment is repeated n times, a sample is obtained from the space $R^n = \{(x_1, \ldots, x_n) | x_i \in R\}$.

Let

$$F^n = \{A_1 \times A_2 \times \cdots \times A_n\} = \{(x_1, \ldots, x_n) | x_i \in A_i,$$
$$i = 1, \ldots, r\},$$

where $A_i \in F$.

If p is a probability measure on F, there is a unique probability measure on F^n assigning probability

$$p(A_1)p(A_2) \cdots p(A_n)$$

to $A = \prod_{i=1}^{n} A_i$, $A_i \in F$; it is called the *product probability measure*.

[1] See Fraser [10, pp. 139–142].

If p is a probability measure over (R^n, F^n), then p is symmetric. That is, $p(A) = p(B)$ if $A \in F^n$ and if $B = \{[x_{i_1}, \ldots, x_{i_n}] | (x_1, \ldots, x_n) \in A\}$, where $(i_1, \ldots, i_n)$ is a permutation of $(1, \ldots, n)$.

General order statistic

Definition

$t(x_1, \ldots, x_n) = \{(x_{i_1}, \ldots, x_{i_n}) | (i_1, \ldots, i_n),$ a permutation of $(1, \ldots, n)\}$.

For example, if $n = 3$, $x_1 = x_2 = x_3$, then $t(x_1, x_2, x_3) = \{(x_1, x_2, x_3)\}$. If $x_1 = x_2 \neq x_3$,

$$t(x_1, x_2, x_3) = \{(x_1, x_1, x_3), (x_1, x_3, x_1), (x_3, x_1, x_1)\}$$

is the order statistic.

Lemma 3

If $f(x_1, \ldots, x_n)$ is a symmetric function, it can be written as a function of the order statistic

$$f(x_1, \ldots, x_n) = h[t(x_1, \ldots, x_n)]$$

and conversely, if $f(x_1, \ldots, x_n)$ can be written as a function of the order statistic, then it is a symmetric function.

Proof:

Since $f(x_1, \ldots, x_n)$ is constant-valued for the $n!$ points, $(x_{i_1}, \ldots, x_{i_n})$ correspond to the $n!$ permutations.

Theorem 3

Let $\{p_\theta | \theta \in \Omega\}$ be the family of all symmetric probability measures over (R^n, A^n). Then the general order statistic is sufficient for $(p_\theta | \theta \in \Omega)$.

Proof

Let $A \in R^n$ and let $p(A | \{x_1, \ldots, x_n\})$ stand for the conditional probability of an event falling in the set A, given the order statistic $\{x_1, \ldots, x_n\}$.

We shall show that

$$p(A|\{x_1, \ldots, x_n\}) = \frac{i(A, \{x_1, \ldots, x_n\})}{n!}$$

where $i(A, \{x_1, \ldots, x_n\})$ is defined to be the number of the $n!$ permutations of $(x_{i_1}, \ldots, x_{i_n})$ that fall in A. Let B be a symmetric set in R^n standing for a set of values of the statistic $t(x_1, \ldots, x_n)$. Examining the definition of conditional probability, we find that we need only show that the following equation holds for all B:

$$p_\theta{}^x(A \cap B) = \int_B \frac{i(A, \{x_1, \ldots, x_n\})}{n!} \prod_{i=1}^{n} dp_\theta(x_i) \tag{5}$$

The left-hand side of this equation can be written

$$\int_B \phi_A(x_1, \ldots, x_n) \prod_{i=1}^{n} dp_\theta(x_i)$$

where $\phi_A(x_1, \ldots, x_n)$ is the indicator function of the set A and is defined by

$$\phi_A(x_1, \ldots, x_n) = \begin{cases} 1 & (x_1, \ldots, x_n) \in A \\ 0 & (x_1, \ldots, x_n) \notin A \end{cases}$$

Since the probability measure over R^n is symmetric in the coordinates and so also is the set B, we have that the left-hand side is equal to

$$\int_B \phi_A(x_{i_1}, \ldots, x_{i_n}) \prod_{i=1}^{n} dp_\theta(x_i)$$

where $(i_1, \ldots, i_n)$ is any permutation of $(1, 2, \ldots, n)$. Therefore, we can write the left-hand side as the average of the $n!$ equal expressions obtained by taking the $n!$ permutations of $(1, \ldots, n)$:

$$\int_B \frac{\sum_p \phi_A(x_{i_1}, \ldots, x_{i_n})}{n!} \prod_{i=1}^{n} dp_\theta(x_i)$$

It is easy to see that $\sum_p \phi_A(x_{i_1}, \ldots, x_{i_n})$ is just the number of permutations of $(x_1, \ldots, x_n)$ that fall in A, that is, is equal to $i(A, \{x_1, \ldots, x_n\})$. This establishes equality (5).

Examining the expression for conditional probability $p(A|\{x_1, \ldots, x_n\})$, we see that it does not depend on θ; therefore

$$t(x_1, \ldots, x_n) = \{x_1, \ldots, x_n\}$$

is a sufficient statistic for $\{p_\theta | \theta \in \Omega\}$.

5 □ MINIMAL SUFFICIENT STATISTIC

Definition

t is a minimal sufficient statistic for $P = \{p_\theta | \theta \in \omega\}$ if t has the following properties:

(i) t is a sufficient statistic for $\{p_\theta | \theta \in \omega\}$.
(ii) If t^* is a sufficient statistic for $\{p_\theta | \theta \in \omega\}$, then there is a function s such that $t(x) = s[t^*(x)]$, a.e. p_θ, for each $\theta \in \omega$.

Existence and construction of minimal sufficient statistic[1]

Let the probability measures in $P = \{p_\theta | \theta \in \omega\}$ all be absolutely continuous with respect to some measure μ on F which is independent of θ and has the property that X is a countable union of sets in F of finite measure μ. That is, there exists $p_\theta(x)$ for each $\theta \in \omega$ such that

$$p_\theta(A) = \int_A p_\theta(x) \, du \quad \text{for all } A \in F$$

so that $p_\theta(x)$ is a generalized probability density with respect to μ. One can obtain minimal sufficient statistics for P by applying an operation U to the family $\{p_\theta(x) | \theta \in \omega\}$. Let us begin by defining the operation U and applying the definition to some examples of families of distributions of some statistical interest.

Let $\mathfrak{F}$ denote a family of real-valued functions f on X, and suppose these functions are indexed by a subscript θ taking on values in Λ, so that $\mathfrak{F} = \{f_\theta | \theta \in \Lambda\}$. The result of the operation U on $\mathfrak{F}$ is a decomposition of X, to be denoted by $U(\mathfrak{F})$. For any point x^0 in R, the element D of $U(\mathfrak{F})$ containing x^0, written $D(x^0)$, is defined as the set of all points x

---
[1] See also Lindgren [16, pp. 198–280] for an elementary treatment.

for which there exists a function $k(x,x^0) \neq 0$, not depending on θ, and such that $f_\theta(x) = k(x,x^0)f_\theta(x^0)$ for all $\theta \in \Lambda$. Let us note that if x' is in $D(x^0)$, then x^0 is in $D(x')$; also, that

$$D^0 = \{x | f_\theta(x) = 0 \text{ for all } \theta \in \Lambda\}$$

is an element of $U(\mathfrak{F})$.

Now let us consider examples of applying the operation U. In each case R is a euclidean n-space, and F may be taken to be the family of Borel sets. In all except the first example, μ will be taken as the Lebesgue measure on F.

EXAMPLE 1

Suppose $\mathbf{x} = (x_1, \ldots, x_n)$ is a random sample from a binomial population with parameter θ, x_i taking on the values 1 and 0 with respective probabilities θ and $1 - \theta$. Let μ assign measure 1 to each of the 2^n points in the set R_+ consisting of the points $(x_1, x_2, \ldots, x_n)$ with $x_i = 0$ or 1, and measure zero to $R - R_+$. In this example we might take F as the family of all subsets of X_+ or the family of all subsets of X, instead of the usual family of all Borel sets in W. For ω we take the open interval $0 < \theta < 1$, and for $p_\theta(\mathbf{X})$ the determinations

$$p_\theta(\mathbf{x}) = \begin{cases} \prod_{i=1}^{n} \theta^{x_i}(1-\theta)^{1-x_i} & \text{if } \mathbf{x} \in R_+ \\ 0 & \text{otherwise} \end{cases} \quad (1)$$

For this specification, $D^0 = R - R_+$. For x^0 not in D^0, $\mathbf{x}$ is in $D(\mathbf{x}^0)$ if and only if there exists $k(\mathbf{x},\mathbf{x}^0)$ such that $p_\theta(\mathbf{x}) = k(\mathbf{x},\mathbf{x}^0)p_\theta(\mathbf{x}^0)$, and hence

$$\left(\frac{\theta}{1-\theta}\right)^T = k(\mathbf{x},\mathbf{x}^0) \quad \text{where } T = \sum_1^n x_i - \sum_1^n x_i^0 \quad (2)$$

The left member of (2) is independent of θ if and only if

$$\sum_1^n x_i = \sum_1^n x_i^0 \quad (3)$$

The decomposition is associated with the statistic Σx_i almost everywhere or any one-to-one function of this. For any $\mathbf{x}^0$ in R_+ the element $D(\mathbf{x}^0)$ consists of the $\sum_1^n x_i^0$ points in R_+ satisfying (3). The decomposition contains $n+2$ elements D, namely, D^0, and $n+1$ sets D, in which $\sum_1^n x_i = \nu$, $\nu = 0, 1, 2, \ldots, n$. The same decomposition is obtained if the operation U is applied to any subset of $\{p_\theta(\mathbf{x})\}$ consisting of two or more elements because (3) is valid for this case also. $t = \sum_1^n x_i$ is a minimal sufficient statistic.

EXAMPLE 2

Let $X_1, X_2, \ldots, X_n$ be independent random variables each one normally distributed with mean θ_1 and variance θ_2. Here μ is the Lebesgue measure, and F is the Borel field generated by Borel sets, D^0 is the empty set, and the determination of probability density $p_\theta(x)$ is

$$\frac{dp_\theta(\mathbf{x})}{d\mu} = p_\theta(\mathbf{x}) = (2\pi\theta_2)^{-n/2} \exp -\frac{1}{2\theta_2} \sum_1^n (x_i - \theta_1)^2 \qquad (4)$$

Since the denominator of the fraction $p_\theta(\mathbf{x})/p_\theta(\mathbf{x}^0)$ cannot vanish, $D(\mathbf{x}^0)$ is the set such that this ratio is independent of θ. This is seen to be the same as the set where

$$-\frac{1}{2\theta_2}\left(\sum_1^n x_i^2 - \sum_1^n x_i^{0^2}\right) + \frac{\theta_1}{\theta_2}\left(\sum_1^n x_i - \sum_1^n x_i^0\right)$$

is independent of (θ_1, θ_2), namely, the set where

$$\sum_1^n x_i^2 = \sum_1^n x_i^{02} \quad \text{and} \quad \sum_1^n x_i = \sum_1^n x_i^0 \qquad (5)$$

Thus the decomposition induced by the operation U is that associated with the statistic $\left(\sum_1^n X_i, \sum_1^n X_i^2\right)$, which is a minimal sufficient statistic.

The minimal sufficiency of the associated statistics considered above follows from the following theorem.

Theorem

Suppose P is a countable set of probability measures on F possessing probability density functions $p_\theta(x)$, $\theta \in \omega$, with respect to μ. If the operation U is applied to a particular determination of the family $\{p_\theta(x)\}$, the resulting decomposition is F-measurable and is associated with a minimal sufficient statistic for P.

For the proof see Lehmann and Scheffé [13].

REMARK

For further work on sufficiency one can refer to [1] to [8].

PROBLEMS

1. Let X be a random variable with hypergeometric density function (n,D,N), where n and N are fixed, $D \in \omega$, where ω is the set of nonnegative integers $\{0, 1, 2, \ldots, N\}$. Prove that X is sufficient for D.

2. Let $X_1, X_2, \ldots, X_n$ be independent random variables each with common density function $N(\mu,\sigma^2)$. Prove that $\{\bar{X}, \Sigma(X_i - \bar{X})^2\}$ is sufficient for $\{(\mu,\sigma^2), +\infty < \mu < \infty, \sigma^2 > 0\}$.

3. Let $X_1, X_2, \ldots, X_n$ be independent random variables with common density function $N(0,\sigma^2)$. Prove that $\sum_1^n X_i^2$ is a sufficient statistic for σ^2, where $\sigma^2 > 0$.

4. Let X_1, X_2, be independent random variables, with $x_i\ b(n,p_i)$, $0 < p_i < 1$, $i = 1, 2$. Prove that (X_1,X_2) is a sufficient statistic for (p_1,p_2).

5. Let X_1, X_2 be independent random variables, x_i is Poisson distributed with parameter m_i, where $m_i > 0$ and $i = 1, 2$. Prove that (X_1,X_2) is a sufficient statistic for (m_1,m_2).

6. Let W be a n-dimensional euclidean space, and let F be all the class of Borel subsets of W. μ is a σ-finite measure on F, ω is a subset of r-dimensional euclidean space and p_θ, $\theta \in \omega$, is a generalized probability-density function with respect to μ such that

$$p_\theta(x) = k(\theta)h(x)\ \exp\ \sum_1^r \theta_i t_i(\mathbf{x}) \qquad x \in W$$

where $h, t_1, t_2, \ldots, t_r$ are real-valued Borel measurable functions. Prove that $t = (t_1, \ldots, t_r)$ is a sufficient statistic for $\theta = (\theta_1, \ldots, \theta_r)$.

7. Let $X_1, X_2, \ldots, X_n$ be independent random variables with common

density function
$$p_\theta(x) = exp\,[-x^{2n} + \theta_1 x + \theta_2 x^3 + \cdots + \theta_n x^n]$$
where $\theta = (\theta_1, \ldots, \theta_n)$ is a point in n-dimensional space. Prove that $t(x) = (x_{(1)}, x_{(2)}, \ldots, x_{(n)})$ is a sufficient statistic for θ, where $x_{(i)}$ is the ith-order statistic, $i = 1, 2, \ldots, n$.

8. Let
$$\mathbf{x}_1 = (x_{11}, \ldots, x_{1k})$$
$$\mathbf{x}_2 = (x_{21}, \ldots, x_{2k})$$
$$\cdots\cdots\cdots\cdots\cdots$$
$$\mathbf{x}_n = (x_{n1}, \ldots, x_{nk})$$
be independent, each with density function $f_{\mu,\Lambda}$, where
$$f_{\mu,\Lambda} = \frac{|\Lambda|^{\frac{1}{2}}}{(2\mu)^{k/2}} e^{-\frac{1}{2}(\mathbf{x}-\mu)\Lambda(\mathbf{x}-\mu)'}$$
for each $\mathbf{x} = (x_1, \ldots, x_k)$. Let
$$\omega = \{(\mu,\Lambda)|\mu = (\mu_1, \ldots, \mu_k),\ -\infty < \mu_i < \infty,\ i = 1, \ldots, k\}$$
$$\Lambda = \{(\lambda_{ij}),\ i,j = 1, 2, \ldots, k,\ \text{any positive definite matrix}\}$$
Find a sufficient statistic for (μ, Λ).

9. Consider Prob. 8 when Λ is defined as
$$\Lambda = \begin{bmatrix} \lambda_{11} & & & \\ & \lambda_{22} & & 0 \\ & 0 & \ddots & \\ & & & \lambda_{nn} \end{bmatrix}$$

10. Let $X_1, X_2, \ldots, X_n$ be independent random variables, each with density function $f_{\theta,\eta}(x)$, where
$$f_{\theta,\eta}(x) = \begin{cases} \dfrac{1}{\eta - \theta} & \text{if } \theta < x < \eta \\ 0 & \text{otherwise} \end{cases}$$
for the parameter space $\omega = \{(\theta,\eta)|\theta,\eta\text{ real, with }\theta < \eta\}$. Find a sufficient statistic for (θ,η).

11. Repeat Prob. 10 when $\eta = \theta + 1$.

12. State and prove the Neyman factorization criterion when the density function of a random variable is continuous.

13. (Lehmann and Scheffé) If the family P of probability measures on W possesses a generalized density $p_\theta(x)$ with respect to μ, and if the family $P = \{p_\theta | \theta \in \omega\}$ of densities is separable (μ), then there exists a minimal sufficient statistic for P and it may be constructed by applying the operation U to any countable set p_1 dense (μ) in p. Prove the statement.

14. Let X be a random variable with density function

$$f_\theta(x) = \tfrac{1}{2} \qquad \theta \leq x \leq \theta + 2$$

where θ is any real number. Then prove that there does not exist a sufficient statistic for θ.

15. Suppose that $X_1, \ldots, X_n$ are random variables taking on the values 0 or 1 such that

$$P_{D,N}(x_1 = 1) = \frac{D}{N}$$

$$P_{D,N}\{X_{k+1} = 1 | X_1 = x_1, \ldots, X_k = x_k\} = \frac{D - \sum_{1}^{k} x_i}{N - k}$$

where

$$1 \leq k \leq n-1 \qquad \sum_{i=1}^{k} x_i \leq D \qquad \sum_{i=1}^{n} (1 - x_i) \leq N - D \qquad 1 \leq n \leq N$$

Let

$$\Omega = \{(D,N) | D = 0, 1, \ldots, N,\ N = n, n+1, \ldots\}$$

Prove that $t(\mathbf{x}) = \sum_{i=1}^{n} X_i$ is a sufficient statistic for Ω.

REFERENCES

1 BAHADUR, R. R.: Sufficiency and Statistical Decision Functions, *Ann. Math. Statist.*, vol. 25, pp. 423–462, 1954.

2 BAHADUR, R. R.: Statistics and Subfields, *Ann. Math. Statist.*, vol. 26, pp. 490–497, 1955.

3 BAHADUR, R. R., and E. L. LEHMANN: Two Comments on Sufficiency and Statistical Decision Functions, *Ann. Math. Statist.*, vol. 26, pp. 139–142, 1955.

4 BARANKIN, E. W., and M. KATZ, JR.: Sufficient Statistics of Minimal Dimension, *Sankhyā, ser. A*, vol. 21, pp. 217–246, 1959.

5 BARANKIN, E. W., and A. P. MAITRA: Generalization of the Fisher-Darmois-Koopman-Pitman Theorem on Sufficient Statistics, *Sankhyā, ser. A*, vol. 25, pp. 217–244, 1963.

6 BURKHOLDER, D. L.: Sufficiency in the Undominated Case, *Ann. Math. Statist.*, vol. 32, pp. 1191–1200, 1961.

7 BURKHOLDER, D. L.: On the Order Structure of the Set of Sufficient Subfields, *Ann. Math. Statist.*, vol. 33, pp. 596–599, 1962.

8 FRASER, D. A. S.: Local Conditional Sufficiency, *J. Roy. Statist. Soc., ser. B*, vol. 26, pp. 52–62, 1964.

9 FRASER, D. A. S.: Sufficient Statistics and Selection Depending on the Parameter, *Ann. Math. Statist.*, vol. 23, pp. 417–425, 1952.

10 FRASER, D. A. S.: "Nonparametric Methods in Statistics," John Wiley & Sons, Inc., New York, 1957.

11 HALMOS, P. R., and L. J. SAVAGE: Application of the Radon-Nikodym Theorem to the Theory of Sufficient Statistics, *Ann. Math. Statist.*, vol. 20, pp. 225–241, 1949.

12 KOOPMAN, B. O.: On Distributions Admitting a Sufficient Statistic, *Trans. Am. Math. Soc.*, vol. 39, pp. 399–409, 1956.

13 LEHMANN, E. L., and H. SCHEFFÉ: Completeness, Similar Regions, and Unbiased Estimation, *Sankhyā, ser. A*, pt. 4, vol. 10, pp. 305–340, 1950.

14 NEYMAN, J.: Su un teorema concernente le cosiddette statistiche sufficienti, *Inst. Ital. Atti Giorn.*, vol. 6, pp. 320–334, 1935.

15 PITCHER, T. S.: Sets of Measures Not Admitting Necessary and Sufficient Statistics or Subfields, *Ann. Math. Statist.*, vol. 28, pp. 267–268, 1957.

16 LINDGREN, B. W.: "Statistical Theory," The Macmillan Company, New York, 1960.

chapter Six

Convex Functions

1 □ INTRODUCTION

In estimation problems, one is presented with a loss function and then attempts to find appropriate estimates which minimize the average or expected loss. The loss function is quite commonly a convex function. For this reason, we now discuss convex functions, stating some examples and some properties to be used later. In subsequent chapters, we shall investigate the role of the convex function as a loss function.

Definition

Let I be an interval on the real line. If g is a real-valued function on I such that, for each $\alpha \in (0,1)$ and each pair (x_1,x_2) of elements of I,

$$g[\alpha x_1 + (1 - \alpha)x_2] \leq \alpha g(x_1) + (1 - \alpha)g(x_2) \qquad (1)$$

then g is convex. The function g is strictly convex if

$$g[\alpha x_1 + (1-\alpha)x_2] < \alpha g(x_1) + (1-\alpha)g(x_2) \tag{2}$$

for each $\alpha \in (0,1)$, $x_1 \in I$ and $x_2 \in I$ with $x_1 \neq x_2$. Note that

$$g(x) = g\left(\frac{x_2 - x}{x_2 - x_1}x_1 + \frac{x - x_1}{x_2 - x_1}x_2\right) \leq \frac{x_2 - x}{x_2 - x_1}g(x_1) + \frac{x - x_1}{x_2 - x_1}g(x_2)$$

if x is between x_1 and x_2.

2 □ EXAMPLES OF CONVEX FUNCTIONS

One can easily check convexity of the following functions:

Function	Interval of Definition
$g(x) = x^2$	I the real line
$g(x) = \|x - a\|^p$	$p \geq 1$
	a any real number
	I the real line
$g(x) = e^x$	I the real line
$g(x) = -\log x$	$I = \{x \mid 0 < x < \infty\}$
$g(x) = 7$	I the real line
$g(x) = \dfrac{1}{x}$	$I = \{x \mid 0 < x < \infty\}$
$g(x) = \begin{cases} 0 \\ 1 \end{cases}$	$0 \leq x < 1$
	$x = 1$
	I the interval $[0,1]$

Now we shall give sufficient conditions for the convexity of a function.

3 □ SUFFICIENT CONDITIONS FOR CONVEXITY

Theorem

Let a function g be defined and twice differentiable throughout the closed interval $[c,d]$, for arbitrary real numbers c and d, with $c < d$. Then g is convex on $[c,d]$ if $g'' \geq 0$ on $[c,d]$.

Proof:

Since for any two points a and b in $[c,d]$ with $a < b$, the mean-value theorem states that

$$g(b) - g(a) = g'(\eta)(b - a) \qquad a < \eta < b \qquad (1)$$

we have, for $c \leq x_1 < x_2 \leq d$ and $0 < \alpha < 1$,

$$g[\alpha x_1 + (1 - \alpha)x_2] - g(x_1) = (1 - \alpha)(x_2 - x_1)g'(\eta_1) \qquad (2)$$

where

$$x_1 < \eta_1 < \alpha x_1 + (1 - \alpha)x_2$$

and

$$g(x_2) - g[\alpha x_1 + (1 - \alpha)x_2] = \alpha(x_2 - x_1)g'(\eta_2) \qquad (3)$$

where

$$\alpha x_1 + (1 - \alpha)x_2 < \eta_2 < x_2$$

Solving (2) and (3) for $g'(\eta_1)$ and $g'(\eta_2)$ and noting that $g'(\eta_1) \leq g'(\eta_2)$, since $g''(x) \geq 0$, we obtain

$$g[\alpha x_1 + (1 - \alpha)x_2] \leq \alpha g(x_1) + (1 - \alpha)g(x_2) \qquad (4)$$

which proves the theorem.

4 □ PROPERTIES OF CONVEX FUNCTIONS

Theorem 1

If g is a convex function on the real interval I, and if x_0 is an interior point of I, then there is a real number m such that

$$g(x) \geq g(x_0) + m(x - x_0) \qquad x \in I$$

If g is strictly convex, the inequality is strict for $x \in I$, $x \neq x_0$.

Proof:

The theorem is a direct consequence of the following two statements:

(i) $f(x) = \dfrac{g(x) - g(x_0)}{x - x_0}$ is a nondecreasing (increasing if g is strictly convex) function of x for $x \in \{I - x_0\}$.

(ii) For

$$f(x - 0) = \lim_{\epsilon \to 0} f(x - \epsilon) \qquad \text{and} \qquad f(x + 0) = \lim_{\epsilon \to 0} f(x + \epsilon)$$

for $\epsilon > 0$
defined, $f(x - 0) \leq f(x + 0)$.

For m such that $f(x - 0) \leq m \leq f(x_0 + 0)$, m has the desired property, since if $x < x_0$, $x \in I$, then

$$f(x) \leq f(x - 0) \leq m$$
$$\frac{g(x) - g(x_0)}{x - x_0} \leq m$$

therefore

$$g(x) \geq g(x_0) + m(x - x_0)$$

because $x - x_0$ is negative.

Proof of (i):

Let

$$x_1 \in \{I - x_0\}$$
$$x_2 \in \{I - x_0\}$$
$$x_1 < x_2$$

to show that $f(x_1) \leq f(x_2)$.

We shall consider the following three cases:

Case A:

$$x_1 < x_2 < x_0$$

Case B:

$$x_1 < x_0 < x_2$$

Case C:

$$x_0 < x_1 < x_2$$

Let us consider case C.

We must show that $f(x_1) \leq f(x_2)$ or, equivalently, that

$$\frac{g(x_1) - g(x_0)}{x_1 - x_0} \leq \frac{g(x_2) - g(x_0)}{x_2 - x_0}$$

or that

$$g(x_1) \leq \frac{x_1 - x_0}{x_2 - x_0} g(x_2) + g(x_0)\left(1 - \frac{x_1 - x_0}{x_2 - x_0}\right)$$

$$\leq \frac{x_1 - x_0}{x_2 - x_0} g(x_2) + \frac{x_2 - x_1}{x_2 - x_0} g(x_0)$$

Since the last inequality is true due to the convexity of g, it implies the first inequality, $f(x_1) \leq f(x_2)$ and one can prove the result similarly for cases A and B also.

Similarly one can prove the inequality $f(x - 0) \leq f(x + 0)$ in (ii).

Theorem 2 (Jensen's Inequality)

If g is a convex function on the real interval I, X is a random variable with finite expectation, and $p(X \in I) = 1$, then $E[g(X)] \geq g[E(X)]$.

If g is strictly convex, then $E[g(X)] > g[E(X)]$ unless x has all its probability at one point, and then $E[g(X)] = g[E(X)]$. If x has probability at, say, two points,

$$p(X = x_1) = \alpha$$
$$p(X = x_2) = 1 - \alpha$$

Case A:

Suppose that $E(X)$ is a boundary point. Then $p[X = E(X)] = 1$, and so

$$E[g(X)] = g[E(X)]$$

Proof:

Suppose that $I = [a, \ )$ (or $[a, \ ]$) and that $E(X) = a$. Then

$$p\{X - a > \epsilon\} \leq \frac{E(X) - a}{\epsilon} = 0 \quad \text{for } \epsilon > 0$$

Therefore
$$E[g(X)] = g(a)$$
but
$$g[E(X)] = g(a)$$

and the proof is immediate.

Case B:

Suppose that $E(X)$ is an interior point of I. Then there is a real number m such that

$$g(X) \geq g[E(X)] + m[X - E(X)] \quad X \in I$$

Since $g(X) \geq g(E[X]) + m(X - E[X])$ with probability 1, then

$$\begin{aligned} E[g(X)] &\geq E\{g[E(X)] + m[X - E(X)]\} \\ &= g[E(X)] + m[E(X) - E(X)] \\ &= g[E(X)] \end{aligned}$$

If $p[X \neq E(X)] > 0$ and g is strictly convex, then
$$g(X) > g[E(X)] + m[X - E(X)]$$
with positive probability. Hence
$$E[g(X)] > g[E(X)]$$
and the proof is complete.

EXAMPLE

Let X be a random variable with finite expectation and I the real line unless otherwise stated.

(i) $E(X^2) \geq [E(X)]^2$.
$E(X^2) > [E(X)]^2$ if $p[X \neq E(X)] > 0$.

(ii) $E|X|^p \geq |E(X)|^p$ $p \geq 1$, since $g(X) = |X|^p$ is convex.

(iii) $(E|X|^r)^{1/r} \leq (E|X|^s)^{1/s}$, $0 < r < s$, a form of Hölder's inequality.

Proof:
$$(E|X|^p)^{1/p} \geq E|X| \quad \text{for } p > 1$$
Putting $y = |x|^r$, $p = s/r$,
$$(E|y|^{r \times s/r})^{r/s} \geq E|y|^r$$
and so $(E|y|^s)^{1/s} \geq (E|y|^r)^{1/r}$ and the result follows with a change in notation.

(iv) $E(e^{tX}) \geq e^{tE(X)}$ for t real.

(v) $\log[1/E(1/X)] \leq E(\log X) \geq \log[E(X)]$ when $p(X > 0) = 1$.

Proof:
$g(x) = -\log x$ is convex, and so $E(-\log X) \geq -\log E(X)$. Therefore
$$E(\log X) \leq \log[E(X)].$$
Similarly
$$E\left(\log \frac{1}{X}\right) \leq \log \frac{1}{E(1/X)}$$
and the result follows.

(vi) $E(1/X) \geq \dfrac{1}{E(X)}$ if $p(X > 0) = 1$.

PROBLEMS

1. Let I be a real interval, g be continuous on I and be such that if $x_1 \in I$ and $x_2 \in I$, then
$$g\left(\frac{x_1 + x_2}{2}\right) \leq \frac{g(x_1) + g(x_2)}{2}$$
Prove that g is convex.

2. If g is convex on I, prove that g is continuous at each interior point of I.

3. Let g be a convex function on the set of real numbers. Show that if g is not constant, then g is unbounded.

REFERENCES

1 BLACKWELL, D., and M. A. GIRSHICK: "Theory of Games and Statistical Decision," John Wiley & Sons, Inc., New York, 1954.

2 HARDY, G., J. LITTLEWOOD, and G. POLYA: "Inequalities," Cambridge University Press, New York, 1952.

chapter Seven

Unbiased Estimation

1 □ INTRODUCTION

In this chapter we define mean-unbiased estimation and give examples which throw light on the structure of the estimate and its performance. Then we look for locally best unbiased estimates. Among the class of unbiased estimates we pick up one which gives uniformly minimum variance. The various methods of construction of the estimate are given, the methods based on complete, sufficient statistics are emphasized, and the methods of transforms to construct mean-unbiased estimates are given in Probs. 5 and 6.

A median-unbiased estimate is also defined, and its main property with respect to absolute error is singled out. In Sec. 7 a modal-unbiased estimate is defined. Its various properties are investigated, and various comparisons with respect to performance risk of mean, median, and modal-unbiased estimates are discussed. Some of the results are published for the first time.

The material on unbiased estimation is vast. It is practically impossible to pin down each and every result. Patil's [11] unbiased estimation by additive number theory is very interesting from the point of view of constructing an unbiased estimate. The work of Birnbaum [4] is interesting from the point of view of utility and motivation.

2 □ SOME DEFINITIONS

Definition 1 Unbiasedness

Let X be a random variable with density function $f_\theta(x)$, where $-\infty < \theta < \infty$ and $-\infty < x < \infty$. $\delta(X)$ is said to be an unbiased estimate of θ if $E_\theta[\delta(X)] = \int_{-\infty}^{\infty} \delta(x) f_\theta(x)\, dx = \theta$. The terminology of unbiasedness was introduced by Neyman and David. This condition of unbiasedness prevents an estimate $\delta(X)$ from being too partial to any one value of θ. This principle can be found in work of Gauss and Markoff.

Consider the squared-error loss function, that is, $[\delta(X) - \theta]^2$, this being the loss involved when $\delta(X)$ is an estimate of the true value of θ. Then the risk function is defined as the expectation of the loss function, that is,

$$R_\delta(\theta) = \int_{-\infty}^{\infty} [\delta(x) - \theta]^2 f_\theta(x)\, dx$$

Since $\theta = E_\theta[\delta(X)]$, $R_\delta(\theta)$ is the variance of the estimate.

Definition 2 Uniformly Minimum-variance-unbiased Estimate

Let Δ be a class of all unbiased estimates of θ. If $R_\delta(\theta) < R_{\delta'}(\theta)$ for all θ and for every $\delta' \in \Delta$, then δ is called a *uniformly minimum-variance-unbiased estimate*.

EXAMPLE 1

Let X be a random variable with density function

$$f_\theta(x) = \frac{1}{\sqrt{2\pi}} e^{-(x-\theta)^2/2} \qquad -\infty < x < \infty$$
$$-\infty < \theta < \infty$$

Then $\delta(X) = X$ is a uniformly minimum-variance-unbiased estimate of θ.

Proof:

$$E_\theta[\delta(X)] = \int_{-\infty}^{\infty} \frac{x}{\sqrt{2\pi}} e^{-(x-\theta)^2/2} \, dx = \theta$$

$$\delta'(X) = X + g(X)$$

where $g(X)$ is an unbiased estimate of 0, that is,

$$\int_{-\infty}^{\infty} \frac{g(X)}{\sqrt{2\pi}} e^{-(x-\theta)^2/2} \, dx = 0$$

But since $f_\theta(x)$ is a complete family of density functions,

$$g(X) = 0 \quad \text{a.e.}$$

X is the unique almost everywhere unbiased estimate of θ, and hence it is the uniformly minimum-variance-unbiased estimate of θ.

EXAMPLE 2

Let X be a random variable with $P(X = k) = 1/N$, $k = 1, \ldots, N$. Let $\Omega = \{\text{integers } N | N \geq M\}$, $M > 1$.

(i) Characterize all unbiased estimates.
(ii) Find a uniformly minimum-variance estimate of N.

Case A:

Consider an unbiased estimate of 0; then

$$0 = \frac{1}{N} \sum_{i=1}^{N} g(i) \quad \text{for } N \geq M$$

Therefore $\sum_{i=1}^{M} g(i) = 0$ and $\sum_{i=1}^{N} g(i) = 0$ of all $N > M$. This clearly requires $g(i) = 0$ for $i \geq M + 1$

$$0 = \sum_{i=1}^{M+1} g(i) = g(M+1) + \sum_{i=1}^{M} g(i) = g(M+1)$$

and so on inductively. Thus the class of unbiased estimate of 0 is the set of all functions of positive integers such that

$$g(i) = 0 \quad i \geq M+1 \quad \sum_{i=1}^{M} g(i) = 0$$

Case B:

Let $\gamma: \Omega \to$ reals be a function for which an unbiased estimate of γ is to be found. Thus for $N \geq M$, $N_\gamma(N) = \sum_{i=1}^{N} f(i)$, where $f(i)$ is to be determined. Define $\gamma(i) = 0$ for $1 \leq i < M$ and observe that f is uniquely determined as $f(1) = \gamma(1) = 0$

$$f(i) = i[\gamma(i)] - (i-1)\gamma(i-1) \qquad i > 1$$

for then

$$\sum_{i=1}^{k} f(i) = \sum_{i=1}^{k} i[\gamma(i)] - \sum_{i=1}^{k} (i-1)\gamma(i-1) = \sum_{i=1}^{k} i\gamma(i)$$
$$- \sum_{i=1}^{k-1} i\gamma(i) = k\gamma(i)$$

By the choice of $\gamma(i)$, for $i < M$ we have

$$f(i) = 0 \qquad i < M$$
$$f(M) = M\gamma(M)$$
$$f(i) = i\gamma(i) - (i-1)\gamma(i-1) \qquad \text{for } i > M$$

Case C:

Every unbiased estimate of γ is then $f + g$, where g is an unbiased estimate of 0. Thus suppose that g is an unbiased estimate of zero.

Then

$$E(f + g - \gamma)^2 = E(f - \gamma)^2 + 2E[(f - \gamma)g] + E(g^2)$$

But

$$E[g(f - \gamma)] = \frac{1}{N} \sum_{i=1}^{N} g(i)[i\gamma(i) - (i-1)\gamma(i-1) - \gamma(N)]$$
$$= \frac{1}{N} \sum_{i=1}^{M} g(i)[i\gamma(i) - (i-1)\gamma(i-1) - \gamma(N)]$$
$$= g(M) \frac{M\gamma(M)}{N} - \frac{1}{N} \sum_{1}^{M} g(i)\gamma(N)$$
$$= \frac{M}{N} g(M)\gamma(M)$$

$$\sum_{1}^{M-1} g(i) = 0 \text{ and } \gamma(i) = 0 \text{ for } i < M$$

Thus

$$2E[g(f - \gamma)] + E(g^2) = \frac{1}{N}[g^2(1) + \cdots + g^2(M) + 2M\gamma(M)g(M)]$$

Let $a = M\gamma(M)$ so that

$$\text{LHS} = \frac{1}{N}[g^2(1) + \cdots + (gM + a)^2 - a^2]$$

where LHS stands for left-hand side.

As is well known, $\sum_{i=1}^{n} y_i^2$ subject to $\sum_{1}^{n} y_i = c$ is minimized by $y_i = c/n$. Hence, $g(i) = a/M$, $i < M$,

$$g(M) = \frac{a}{M} - a = -\frac{(M-1)a}{M}$$

Therefore, the minimum-variance estimate f^* is defined by

$$f^*(i) = \frac{a}{M} \qquad i = 1, \ldots, M-1$$

$$f^*(M) = M\gamma(M) - \frac{M-1}{M}a = M\gamma(M) - (M-1)\gamma(M) = \gamma(M)$$

$$f^*(i) = i\gamma(i) - (i-1)\gamma(i-1) \qquad \text{for } i > M$$

Case D:

For a general loss function $W(d,N)$, we have

$$R_\delta = \frac{1}{N} \sum w[f(i) + g(i), N]$$

$$= \frac{1}{N}\left\{\sum_{i=1}^{M-1} w[g(i), N] + w[M\gamma(M) + g(M), N] + \frac{1}{N}\sum_{i=M+1}^{N} W[f(i), N]\right\}$$

An unbiased estimate may be absurd.

EXAMPLE 3

Let X be a random variable with distribution

$$P(X = x) = e^{-\lambda}\frac{\lambda^x}{x!} \qquad x = 0, 1, 2, \ldots$$

Find an unbiased estimate of $\gamma(\lambda) = e^{-3\lambda}$. We prove that

$$\delta(X) = (-2)^X$$

is an unbiased estimate.

Proof:

$$E_\lambda \delta = \sum_{x=0}^{\infty} \frac{(-2)^x}{x!} e^{-\lambda} \lambda^x = e^{-\lambda} \sum \frac{(-2\lambda)^x}{x!} = e^{-3\lambda}$$

$\gamma(\lambda)$ is positive, and the above estimate is sometimes negative, which is absurd.

An unbiased estimate may not exist.

EXAMPLE 4

Let X be a random variable distributed as

$$P_N(X = k) = \frac{\binom{D}{k}\binom{N-D}{n-k}}{\binom{N}{n}} \quad \begin{array}{l} k = 0, 1, \ldots, n \\ N = D, D+1, \ldots \end{array}$$

Is there an unbiased estimate of N?

There is no unbiased estimate of N. The sample size does not change and consequently the set of points $(0, 1, \ldots, n)$ having nonzero probability does not change either. Therefore, for any estimate f,

$$\min_{0 \le k \le n} f(k) \le E(f) \le \max_{0 \le k \le n} f(k)$$

If $f(k)$ is an unbiased estimate of N, then

$$\lim_{N \to \infty} E(f) \le \max_{0 \le k \le n} f(k)$$

which is a contradiction.

3 □ LOCALLY BEST UNBIASED ESTIMATE

Let X be a random variable with density function $P_\theta(x)$, where $\theta \in \Omega$. Let $\mathfrak{M}$ be the class of unbiased estimates of θ. Let the loss function be the squared error.

Definition

$\delta \in \mathfrak{M}$ is said to be a locally best unbiased estimate of θ at θ_0 if

$$E[\delta(x) - \theta_0]^2 \le E[\delta^*(x) - \theta_0]^2 < \infty$$

for all $\delta^* \in \mathfrak{M}$. We shall illustrate this by the following example.

EXAMPLE

Let X be a random variable with density function

$$P_\theta(X = \theta) = \tfrac{1}{2}$$
$$P_\theta(X = \theta + 1) = \tfrac{1}{2}$$
$$\mathfrak{X} = \Omega = \text{set of all integers}$$

For any estimate δ, $E_\theta(\delta) = [\delta(\theta) + \delta(\theta+1)]/2$. Let $\gamma(\theta) = \theta + \tfrac{1}{2}$. $\delta(X) = X$ is an unbiased estimate of γ. Suppose δ_0 is an unbiased estimate of 0; then

$$\frac{\delta_0(\theta) + \delta_0(\theta+1)}{2} = 0 \qquad \theta \in \Omega$$

$$\delta_0(\theta + 1) = -\delta_0(\theta) \qquad \theta \in \Omega$$
$$\delta_0(x) = k(-1)^x \qquad k = \delta_0(0)$$

Let Δ be the class of unbiased estimates of γ; then $\delta \in \Delta$ if and only if there is a real number k such that

$$\delta(x) = x + k(-1)^x \qquad x \in \mathfrak{X}$$

$$\begin{aligned}
\operatorname{var} \delta_k &= E_\theta \delta^2 - (E_\theta \delta)^2 \\
&= \frac{[\theta + k(-1)^\theta]^2 + [\theta + 1 + k(-1)^{\theta+1}]^2}{2} - (\theta + \tfrac{1}{2})^2 \\
&= \frac{\theta^2 + 2\theta k(-1)^\theta + k^2(-1)^{2\theta} + (\theta+1)^2}{2} \\
&\qquad \frac{- 2(\theta+1)k(-1)^\theta + k^2(-1)^{2\theta}}{} \\
&\qquad - (\theta^2 + \theta + \tfrac{1}{4}) \\
&= \theta^2 + \theta + \tfrac{1}{2} + k^2(-1)^{2\theta} - k(-1)^\theta - \theta^2 - \theta - \tfrac{1}{4} \\
&= \tfrac{1}{4} - k(-1)^\theta + k^2(-1)^{2\theta} = [\tfrac{1}{2} - k(-1)^\theta]^2 \\
&\inf_{\delta \in \Delta} \operatorname{var}_\theta \delta = \inf_{-\infty < k < \infty} \operatorname{var}_\theta \delta_k = 0
\end{aligned}$$

For any $\theta_0 \in \Omega$, set $k = (-1)^{\theta_0}/2$.

$$\operatorname{var}_{\theta_0} \frac{\delta(-1)^{-\theta_0}}{2} = 0$$

Suppose $\delta \in \Delta$ and $\theta_0 \in \Omega$. δ is the locally best unbiased estimate at θ_0 if $\operatorname{var}_{\theta_0} \delta = \inf_{\delta^* \in \Delta} \operatorname{var}_{\theta_0} \delta^*$.

For further detailed study on the subject refer to Barankin [1].

4 □ THE USE OF A COMPLETE, SUFFICIENT STATISTIC TO CONSTRUCT MEAN - UNBIASED ESTIMATES

Rao-Blackwell Theorem

Rao and Blackwell independently developed a powerful method of finding minimum-variance-unbiased estimates. Basically, this technique combines a sufficient statistic for some $g(\theta)$ with an unbiased estimate of $g(\theta)$ to produce the desirable estimator.

Theorem 1

Given that a sufficient statistic $t(x)$ exists for $\{P_\theta | \theta \in \Omega\}$ over the space $(\mathfrak{X}_0, \mathfrak{A})$ and that $\delta(X)$ is an unbiased estimator of $g(\theta)$, it may be said that

$$h(t) = E_\theta[\delta(X)|t(X)]$$

is an unbiased estimator of $g(\theta)$ based on $t(X)$. Furthermore, unless

$$\delta(X) = h[t(X)]$$

a.e. P_θ, it may be concluded that

$$\sigma_\delta^2(\theta) > \sigma_h^2(\theta)$$

and

$$R_\delta(\theta) > R_h(\theta)$$

for strictly convex loss functions. If $\delta(X) = h[t(X)]$, $R_\delta(\theta) = R_h(\theta)$.

Proof:

Since t is a sufficient statistic, the conditional probability will be independent of θ and $h(t) = E[\delta(X)|t(X) = t]$ is not a function of θ. Therefore, $h(t)$ is a statistic defined over the space of t, say F. Now investigating $E_\theta(h)$,

$$\begin{aligned} E_\theta[h(t)] &= \int_F h(t)\, dP_\theta^t(t) \\ &= \int_F \int_{\mathfrak{X}_0} \delta(x)\, dp(x|t)\, dp_\theta^t(x) \\ &= \int_{\mathfrak{X}_0} \delta(x)\, dp_\theta(x) = E_\theta[\delta(X)] \end{aligned}$$

Hence, $h(t)$ is an unbiased estimator of $g(\theta)$.

112 PARAMETRIC ESTIMATION

Now, given the sufficient statistic $t(X)$, consider the conditional risk. For $\delta(X)$ the risk is

$$E(E\{W[\delta(X),\theta]|t\})$$

and for $h(t)$ it is

$$E\{W[h(t),\theta]\}$$

Using the Jensen inequality

$$E(E\{W[\delta(X),\theta]|t\}) \geq E\{W[h(t),\theta]\}$$

or

$$R_\delta(\theta) \geq R_h(\theta)$$

with equality if and only if $\delta(X) = h(t)$.

EXAMPLE 1

Let $X_1, X_2, \ldots, X_n$ be independent random variables with uniform distribution over a set of N positive integers, that is,

$$P(X_i = k) = \begin{cases} \dfrac{1}{N} & \text{for } k = 1, 2, \ldots, N \\ 0 & \text{elsewhere} \end{cases}$$

Using the techniques of the Rao-Blackwell theorem, find a uniformly minimum-variance-unbiased estimate of N.

It has been previously found that $t = \max X_i$ is both a complete and sufficient statistic for this distribution.

$$\frac{[t(X)]^{n+1} - [t(X) - 1]^{n+1}}{[t(X)]^n - [t(X) - 1]^n}$$

is uniformly minimum risk and unbiased, and

$$E(X) = \sum_1^N \frac{k}{N} = \frac{1}{N}\frac{N(N+1)}{2} = \frac{N+1}{2}$$

$$E(2X - 1) = N \qquad N \in r \qquad \text{(set of positive integers)}$$

Let

$$\delta(X) = 2X - 1$$

Then

$$P_N[t(y)] = P_N(t_N = y) = P_N[t(X) \leq y] - P_N[t(X) \leq y - 1]$$

$$= \frac{y^n}{N^n} - \frac{(y-1)^n}{N^n}$$

which from previous examples is known to be complete. Then the conditional probability is

$$P_N[X_1 = x_1, \ldots, X_n = x_n | t(\mathbf{x}) = y]$$

$$= \frac{P_N[X_1 = x_1, \ldots, X_n = x_n, t(\mathbf{x}) = y]}{P_N[t(\mathbf{x}) = y]}$$

$$= \frac{1/N^n}{[y^n - (y-1)^n]/N^n}$$

$$P_N\{x_1, x_2, \ldots, x_n | t(x) = y\} = \begin{cases} \dfrac{1}{y^n - (y-1)^n} & \text{if } t(x) = y \\ 0 & \text{otherwise} \end{cases}$$

$$P\{x_1 | t(x) = y\} = \frac{P[x_1, t(x) = y]}{P[t(x) = y]} = \frac{y^{n-1} - (y-1)^{n-1}}{y^n - (y-1)^n}$$

$$\text{for } x_1 = 1, 2, \ldots, y-1$$

$$= \frac{y^{n-1}}{y^n - (y-1)^n} \quad \text{if } x_1 = y$$

$$E(\delta | t)_y = \frac{y^{n-1} - (y-1)^{n-1}}{y^n - (y-1)^n} \sum_{1}^{y-1} (2x - 1) + \frac{y^{n-1}(2y - 1)}{y^n - (y-1)^n}$$

Summing,

$$\frac{[y^{n-1} - (y-1)^{n-1}](y-1)^2 + y^{n-1}(2y-1)}{y^n - (y-1)^n} = \frac{y^{n+1} - (y-1)^{n+1}}{y^n - (y-1)^n}$$

Randomized estimates

Let X be a random variable with density function $P_\theta(x)$. We are interested in finding an estimate of θ, say. Previously, what we did was to construct a function of $X = x$, say δ, with the desired property of the risk and call it an estimate of θ. This type of an estimate is known as a *nonrandomized estimate*.

Definition 1

If $X = x$, we observe a random variable Z_x having a known distribution function depending on x. $Z_x = z$ is taken as an estimate of θ. Z_x is called a randomized estimate of θ.

The following group of theorems will be very useful in finding unbiased estimates in many particular cases and also in ranking the relative desirability of various unbiased estimates.

Theorem 2

For a convex loss function, there exists a nonrandomized unbiased estimator which has risk not larger than that of any randomized unbiased estimator.

Proof:
The proof is simple and is left for the reader to supply.

REMARK
See Prob. 6 of Chap. 10, concerning admissible and minimax estimation.

Theorem 3

If $\{P_\theta | \theta \in r\}$ is a family of symmetric probability measures over $\{\mathcal{K}, \mathcal{C}\}$, t is the general order statistic, and δ is a real-valued estimate which is a function of $x_1, x_2, \ldots, x_n$ through $x_1, x_2, \ldots, x_m$ such that $E_\theta(\delta) < \infty$ for $\theta \in r$, then the estimate $h[t(x_1, x_2, \ldots, x_n)]$ given by the conditional expectation is

$$h[t(x_1, x_2, \ldots, x_n)] = \frac{\sum_{i_1, \ldots, i_m} \delta_0(x_{i_1}, \ldots, x_{i_m})}{n(n-1) \cdots (n-m+1)}$$

$$\text{for } (x_1, x_2, \ldots, x_n) \in R^n$$

is the minimum-variance-unbiased estimate of $E_\theta(\delta)$ among the symmetric estimators.

The indicated sum is over the $n(n-1), \ldots, (n-m+1)$ permutations of $(i_1, \ldots, i_m)$ from $(1, \ldots, n)$, and

$$\delta_0(x_1, x_2, \ldots, x_m) = \delta(x_1, x_2, \ldots, x_n)$$

Proof:

$$\int_B h(y) \, P_\theta^{t^{-1}} = \int_{t^{-1}(B)} h[t(x_1, x_2, \ldots, x_n)] \, dP_\theta(x_1, \ldots, x_n)$$

$$= C \sum_{i_1, \ldots, i_m} \int_{t^{-1}(B)} \delta_0(x_{i_1}, x_{i_m}) \, dP_\theta(x_{i_1}, \ldots, x_{i_m})$$

$$\text{where } C = [n(n-1), \ldots, (n-m+1)]^{-1}$$

$$= \int_{t^{-1}(B)} \delta(x_1, x_2, \ldots, x_n) \, dP_\theta(x_1, \ldots, x_n)$$

Definition 2

An estimable parameter is one for which an unbiased estimator exists. To be precise, a parameter $g(\theta)$ for $\{P_\theta | \theta \in \Omega\}$ is said to be estimable if an unbiased estimator $\delta(X)$ exists for it.

The following two theorems due to Lehmann and Scheffé give some of the properties of estimable parameters.

Theorem 4

All estimable real parameters $g(\theta)$ have a unique minimum-variance-minimum-risk-unbiased estimate if there exists a complete sufficient statistic $t(X)$ for $\{P_\theta | \theta \in \Omega\}$. No other unbiased estimator is a function of the complete sufficient statistic $t(X)$ and loss function is strictly convex.

Proof:

The statement of the theorem assumes that at least one unbiased estimate of $g(\theta)$ exists, and the Rao-Blackwell theorem says that for any unbiased estimator $\delta(X)$ there exists an unbiased estimator $h[t(X)]$ such that $R_\delta(\theta) \geq R_h(\theta)$ with equality if and only if $\delta(X) = h[t(X)]$, a.e. $P_\theta{}^t$. Therefore, only unbiased estimators based on $t(X)$ need be considered.

If two such estimators h_1 and h_2 are assumed,

$$E_\theta[h_1(t)] = g(\theta) \quad \text{and} \quad E[h_2(t)] = g(\theta)$$

which imply that $E[h_1(t) - h_2(t)] = 0$. However, since $t(X)$ is a complete statistic, $h_1(t) - h_2(t) = 0$, a.e. $P_\theta{}^t$. Thus, the estimator based on $t(X)$ is unique almost everywhere, has smaller risk, and has smaller variance than any other unbiased estimator.

Theorem 5

If for $\{P_\theta | \theta \in \Omega\}$ there exists a complete sufficient $t(\mathbf{x})$, then every vector estimable parameter has a unique unbiased estimator with a minimum risk of strictly convex loss function, which is the only unbiased estimate that is a function of the complete sufficient statistic $t(\mathbf{x})$.

Proof:

The proof is left to the reader.

EXAMPLE 2

Consider the independent random variables $X_1, X_2, \ldots, X_n$, where each has distribution $N(\mu,1)$. Then $\bar{X} = \sum_i \dfrac{X_i}{n}$ is a complete and sufficient statistic. Obviously $E(\bar{X}) = \mu$. Now consider the estimate of μ^2:

$$E_\mu(\bar{X}^2) = E[\bar{X} - E(\bar{X})]^2 + E^2(\bar{X}) = \frac{1}{n} + \mu^2$$

Therefore $\bar{X}^2 - 1/n$ is an unbiased estimate of μ^2, and it is uniformly minimum variance for convex loss functions.

The relationship between two functions which are unbiased estimates of the same estimable parameter is what one would expect; i.e., they are equal and it is proved in the following theorem.

Theorem 6

If $\delta_1(X)$ and $\delta_2(X)$ are unbiased estimates of θ having uniformly minimum risk (strictly convex loss), then $\delta_1(X) = \delta_2(X)$, a.e. P_θ.

Proof:

Let $R(\theta)$ be the minimum value of the risk at θ. Then

$$\begin{aligned}R(\theta) &= E_\theta\{L[\delta_1(X),\theta]\} \\ &= E_\theta\{L[\delta_2(X),\theta]\}\end{aligned}$$

Since $\delta(X)_1$ and $\delta_2(X)$ are unbiased, $\alpha\delta_1(X) + (1 - \alpha)\delta_2(X)$ is also an unbiased estimate $f(\theta)$ and it must have of course a risk at least as large as $R(\theta)$, that is,

$$E_\theta\{L[\alpha\delta_1(X) + (1 - \alpha)\delta_2(X)],\theta\} \geq R(\theta) \qquad (1)$$

However, with $\alpha \in (0,1)$ the strict convexity of L gives

$$\begin{aligned}E_\theta\{L[\alpha\delta_1(X) + (1 - \alpha)\delta_2(X)],\theta\} &\leq E_\theta\{\alpha L[\delta_1(X),\theta] + (1 - \alpha)L[\delta_2(X),\theta]\} \\ &\leq \alpha R(\theta) + (1 - \alpha)R(\theta) = R(\theta) \qquad (2)\end{aligned}$$

but with equality if and only if

$$L[\alpha\delta_1(x) + (1 - \alpha)\delta_2(x),\theta] = \alpha L[\delta_1(x),\theta] + (1 - \alpha)L[\delta_2(x),\theta]$$

with probability 1 (P_θ). Again, by strict convexity, this last condition can only hold if $\delta_1(X) = \delta_2(X)$ with probability 1. The two inequalities

(1) and (2) together imply that

$$E_\theta\{L[\alpha\delta_1)(X) + (1 - \alpha)\delta_2(X),\theta]\} = R(\theta)$$

and this equality by the last remark in the above paragraph implies that $\delta_1(X) = \delta_2(X)$ with probability 1. Thus, any two minimum-risk estimators for strictly convex loss are essentially equivalent.

Method of covariance

We now define two important classes of unbiased estimates. First, the class N_0 of unbiased estimates of zero for $\{P_\theta | \theta \in \Omega\}$ which are based on the sufficient statistic $t(x)$. Thus,

$$N_0 = \{\delta(t) | E_\theta[\delta(t)] = 0, \theta \in \Omega\}$$

A second important class is the class of minimum-variance-unbiased estimates N_1

$$N_1 = \left\{ h(t) \;\middle|\; \begin{array}{ll} E_\theta[h(t)] < \infty & \text{for } \theta \in \Omega \\ E[h(t)\delta(t)] = 0 & \text{for } \delta(t) \in N_0 \text{ and for each} \\ & \theta \text{ such that } \sigma_h{}^2 < \infty \end{array} \right\}$$

Theorem 7

If the variance of all statistics in the class N_0 is finite, a statistic is a minimum-variance-unbiased estimate of its expected value if and only if it belongs to the class N_1.

Proof:

The Rao-Blackwell theorem permits us to limit ourselves to estimates based on sufficient statistics $t(x)$. If $\delta(t)$ belongs to the class N_0 and $h(t)$ is a minimum-variance-unbiased estimate of $g(\theta)$, then

$$\begin{aligned} E[h(t) + \lambda\delta(t)] &= E_\theta[h(t)] + \lambda E_\theta[\delta(t)] \\ &= g(\theta) + \lambda 0 = g(\theta) \end{aligned}$$

and $h(t) + \lambda\delta(t)$ is an unbiased estimate of $g(\theta)$. Since $h(t)$ is minimum variance, it is obvious that $h(t) + \lambda\delta(t)$ must have variance which is greater than or equal to the variance of $h(t)$. Hence

$$\sigma^2_{h+\lambda\delta}(\theta) = \sigma_h{}^2(\theta) + \lambda^2\sigma_\delta{}^2(\theta) + 2\lambda E(h\delta) \geq \sigma_h{}^2(\theta)$$

Clearly, if $\sigma_h^2(\theta) < \infty$,

$$\lambda^2 \sigma_\delta^2(\theta) + 2\lambda E_\theta(h\delta) \geq 0$$

for all λ. However, if λ is small enough, $\lambda^2 \sigma_f^2(\theta)$ becomes negligible with respect to $2\lambda E(h\delta)$. Then

$$2\lambda E(h\delta) \geq 0$$

but since λ is arbitrary, $E(h\delta)$ cannot be different from zero, as changing the sign of λ reverses the inequality. Therefore

$$E[h(t)\delta(t)] = 0$$

whenever $\sigma_h^2(\theta) < \infty$, and therefore $h(t)$ belongs to N_1.

To prove the converse, assume that $h(t)$ belongs to N_1 and $h'(t)$ is some other unbiased estimator of $g(\theta)$. Then

$$E[h(t) - h'(t)] = 0$$

Hence $\delta(t) = h(t) - h'(t)$ belongs to N_0. Now it may be said that

$$\sigma_{h'}^2(\theta) = \sigma_{h+\delta}^2(\theta)$$

$$\sigma_{h'}^2(\theta) = \begin{cases} \sigma_h^2 & \text{if } \sigma_h^2 = \infty \\ \sigma_h^2 + \sigma_\delta^2 & \text{if } \sigma_h^2 < \infty \end{cases}$$

which implies that $\sigma_h^2 \leq \sigma_{h'}^2$. And so h is a minimum-variance-unbiased estimate of its expected value.

5 □ MEDIAN UNBIASEDNESS

The previous discussion of unbiased estimation used the mean and the variance of the estimator in order to establish a criterion of goodness of the estimator. In many instances, however, these statistics do not give a true picture of the underlying distribution. The concept of median unbiasedness to be developed in this section is often a more desirable form of estimation than the unbiased estimation.

The median of a distribution $p(x)$ of a random variable is defined in the following way. Let

$$m_1 = \inf_{r \in R^1} [r, P(x \leq r) \geq \tfrac{1}{2}]$$

$$m_2 = \sup_{r \in R^1} [r, P(x \geq r) \geq \tfrac{1}{2}]$$

where R^1 is the real line. Then any $m \in [m_1, m_2]$ is defined as the median of $p(x)$. Stated in another way, the median m of a distribution $p(x)$ is a real number m such that $p(X \geq m) \geq \frac{1}{2}$ and $p(X \leq m) \geq \frac{1}{2}$. Obviously, from this definition, for all real numbers a_1, a_2 such that $m \leq a_1 \leq a_2$ or $m \geq a_1 \geq a_2$

$$E(|X - a_1|) \leq E(|X - a_2|)$$

Let the median of an estimate $\delta(X)$ of $g(\theta)$ belong to the interval (m_1, m_2), and let Ω be connected and $g(\theta)$ be a continuous nonconstant function in a subset of Ω. An estimate of $g(\theta)$ is median unbiased if $g(\theta)$ is the median of $\delta(X)$ for all θ.

Theorem 1

If m is the median of a discrete density $p(x)$ and

$$g(r) = E(|X - r|) = \sum_x |X - r| p(x)$$

then $g(r)$ is minimized for $r = m$ provided that the sum exists for at least one $r \in R^1$.

Proof:

Let $r_0 \in R^1$ and $r_0 \notin [m_1, m_2]$. Now if it can be shown that $E(|X - r_0|) - E(|X - m|) > 0$, it can be concluded that $g(r)$ is minimum for $r = m$. Expanding this expression,

$$\sum_{x \leq m} (r_0 - m) p(x) + \sum_{m < x < r_0} (r_0 + m - 2x) p(x) + \sum_{x > r_0} (m - r_0) p(x)$$

It may be said that

$$\sum_{x \leq m} (r_0 - m) p(x) \geq \tfrac{1}{2}(r_0 - m)$$

from the definition of the median.

$$\sum_{x > r_0} (m - r_0) p(x) = \sum_{x > m} (m - r_0) p(x) - \sum_{m < x \leq r_0} (m - r_0) p(x)$$

Clearly $E(|X - r_0|) - E(|X - m|) > 0$, and therefore the minimum value is attained for $m = r_0$.

It should be added that a similar expression is true for continuous distributions, and the proof is similar.

Theorem 2

Assume the setup of Theorem 1 and if $m \geq a_1 \geq a_2$ or $m \leq a_1 \leq a_2$,

$$E(|X - a_1|) \leq E(|X - a_2|)$$

Since the proof is very simple, it is left to the reader to supply.

6 ☐ MODAL UNBIASEDNESS

In this section we define a modal-unbiased estimate and investigate its properties. We shall compare its risk with that of a mean-unbiased estimate and a median-unbiased estimate.

Definition

Let $X_1, X_2, \ldots, X_n$ be independent random variables with a common density function $f(x;\theta)$, and let $g(X_1, \ldots, X_n)$ be a function of $X_1, X_2, \ldots, X_n$ such that the mode of the density function of $g(X_1, \ldots, X_n)$ is equal to the parameter θ of $f(x,\theta)$; then $g(X_1, \ldots, X_n)$ is said to be a modal unbiased estimate of θ.

EXAMPLE 1

Let $X_1, X_2, \ldots, X_n$ be independent normally distributed with mean θ and variance σ^2. Then $\bar{X} = \left(\sum_{1}^{n} X_i\right)/n$ is normally distributed with mean θ and variance σ^2/n, and the mode of the distribution of $\bar{X}$ is θ; hence $\bar{X}$ is a modal unbiased estimate of θ.

EXAMPLE 2

Let $X_1, X_2, \ldots, X_n$ be independent random variables with density function

$$f(x;\theta) = \begin{cases} \dfrac{1}{\theta} & 0 < x \leq \theta \\ 0 & \text{otherwise} \end{cases}$$

Then $y = \max\limits_{1 \leq i \leq n} X_i$ is a modal-unbiased estimate of θ because the density function of y is

$$\phi(y,\theta) = \frac{ny^{n-1}}{\theta^n} \quad \text{for } 0 < y \leq \theta$$

and the mode of this density function is θ.

REMARK

Let $X_1, X_2, \ldots, X_n$ be independent random variables with a common density function $f(x;\theta)$; then a modal-unbiased estimate of θ may not be unique. This is illustrated by the following example.

EXAMPLE 3

Let $X_1, X_2, \ldots, X_n$ be independent random variables with density function

$$f(x;\theta) = \frac{1}{\theta} e^{-x/\theta} \qquad x > 0$$
$$\theta > 0$$

Let

$$Y_1 = \frac{x_{(n)}}{\log n} \qquad \text{where } x_{(n)} = \max_{1 \leq i \leq n} X_i$$
$$n > 1$$

$$Y_2 = \frac{Z}{n-1} \qquad \text{where } Z = \sum_1^n X_i$$

Y_1 and Y_2 are both modal unbiased estimates of θ.

Proof:

The density function of $x_{(n)}$ is

$$\phi(x_{(n)},\theta) = \frac{n}{\theta} (1 - e^{-x_{(n)}/\theta})^{n-1} e^{-x_{(n)}/\theta} \qquad x_{(n)} > 0$$
$$\theta > 0$$

Since

$$y_1 = \frac{x_{(n)}}{\log n}$$

we have

$$dx_{(n)} = \log n \, dy_1$$

$$\phi(y_1, \theta) = \frac{n \log n}{\theta} (1 - e^{-(y_1/\theta) \log n})^{n-1} e^{-(y_1/\theta) \log n}$$

$$= \frac{n \log n}{\theta} [1 - (n^{-1/\theta})^{y_1}]^{n-1} (n^{-1/\theta})^{y_1} \qquad (1)$$

We have to prove that when $y_1 = \theta$, (1) is maximum, which can be shown easily as follows. Put $n^{-1/\theta} = a$. Then (1) becomes

$$\phi(y_1, \theta) = \frac{n \log n}{\theta} (1 - a^{y_1})^{n-1} a^{y_1} \qquad (2)$$

Differentiate (2) with respect to y and equate it to zero, that is,

$$-\frac{(n-1)n \log n}{\theta}(1-a^{y_1})^{n-2}(a^{y_1} \log a)a^{y_1}$$

$$+\frac{n \log n}{\theta}(1-a^{y_1})^{n-1}a^{y_1} \log a = 0$$

or

$$-(n-1)a^{y_1} + (1-a^{y_1}) = 0 \tag{3}$$

and

$$a^{y_1} = \frac{1}{n} = a^{\theta}$$

Hence $y_1 = \theta$.

It can easily be verified that the second derivative of (2) at $y_1 = \theta$ is negative. Hence (1) is maximum at $y_1 = \theta$, and thus Y_1 is a modal-unbiased estimate of θ.

It is well known that the density function of $Z = \sum_{1}^{n} X_i$ is

$$\psi(z,\theta) = \frac{z^{n-1}}{\Gamma(n)\theta^n} e^{-z/\theta} \quad z > 0$$

$$\theta > 0$$

and so the density function of $Y_2 = Z/(n-1)$ is

$$\psi(y_2,\theta) = \frac{(n-1)^n}{\Gamma(n)\theta^n} y_2^{n-1} e^{-(n-1)y_2/\theta} \quad y_2 > 0$$

$$\theta > 0$$

$$\frac{d\psi(y_2,\theta)}{dy_2} = \frac{(n-1)^{n+1}}{\Gamma(n)\theta^n}(y_2)^{n-2} e^{-(n+1)y_2/\theta} - \frac{(n-1)^{n+1}}{\Gamma(n)\theta^{n-1}} e^{-(n-1)y_2/\theta} = 0$$

that is, $y_2 = \theta$ is the solution and

$$\frac{d^2\psi(y_2,\theta)}{dy_2} = \frac{(n-1)^{n+1}}{\Gamma(n)\theta^n} y_2^{n-3} e^{-(n-1)y_2/\theta} \left[(n-2) - \frac{2(n-1)y_2}{\theta}\right.$$

$$\left. + \frac{n-1}{\theta^2}y_2^2\right]$$

$$\frac{d^2\psi(y_2\theta)}{dy_2}\bigg|_{y_2=\theta} = \frac{(n-1)^{n+1}}{\Gamma(n)\theta^n} \theta^{n-3} e^{-(n-1)}[n-2-2(n-1)+(n-1)] < 0$$

Hence, Y_2 is a modal-unbiased estimate of θ.

Thus a modal-unbiased estimate of a parameter may not be unique.

Theorem 1

Let X be a random variable with density function $f(x;\theta)$, and let $Y = g(X)$. Let $\phi(y)$ be the density function of Y, such that

(i) $\dfrac{d\phi(y)}{dy} = 0$ at $y = \theta$.

(ii) $\dfrac{d^2\phi(y)}{dy^2} < 0$ at $y = \theta$.

(iii) g is a one-to-one transformation from x to y and from y to x.

Then the solution of the following differential equation is a modal unbiased estimate of θ:

(iv) $f(x;\theta) \dfrac{d^2x}{dy^2} + \dfrac{df(x;\theta)}{dx}\left(\dfrac{dx}{dy}\right)^2 = 0$ at $y = \theta$.

(v) $\dfrac{d^2f(x,\theta)}{dx^2}\left(\dfrac{dx}{dy}\right)^3 + 3\dfrac{df(x,\theta)}{dx}\dfrac{dx}{dy}\dfrac{d^2x}{dy^2} + f(x,\theta)\dfrac{d^3x}{dy^3} < 0$.

Proof:

$$\phi(y) = \frac{d}{dy}F(y)$$

$$= \frac{d}{dy}F_x[g^{-1}(y)]$$

$$= \frac{d}{dx}F_x(x)\frac{dx}{dy}$$

$$= f(x,\theta)\frac{dx}{dy}$$

$$\frac{d\phi(y)}{dy} = \frac{df(x,\theta)}{dx}\left(\frac{dx}{dy}\right)^2 + f(x,\theta)\frac{d^2x}{dy^2} = 0$$

at $y = \theta$ by hypothesis and

$$\frac{d^2\phi(y)}{d^2y} = \frac{df(x,\theta)}{dx}\left(\frac{dx}{dy}\right)^3 + 3\frac{df(x,\theta)}{dx}\frac{dx}{dy}\frac{d^2x}{dy^2} + f(x,\theta)\frac{d^3x}{dy^3} < 0$$

at $y = \theta$ by hypothesis; hence the solution of the differential equation is a modal-unbiased estimate of θ.

EXAMPLE 4
Let

$$f(x;\theta) = \frac{1}{\sqrt{2\pi}\,\sigma} e^{-(x-\theta)^2/2\sigma^2}$$

$$\frac{df(x;\theta)}{dx} = \frac{-1}{\sqrt{2\pi}\,\sigma} e^{-(x-\theta)^2/2\sigma^2} \frac{2(x-\theta)}{2\sigma^2}$$

$$\frac{1}{\sqrt{2\pi}\,\sigma} e^{-(x-\theta)^2/2\sigma^2} \frac{d^2x}{dy^2} - \frac{2(x-\theta)}{2\sigma^2} \frac{1}{\sqrt{2\pi}\,\sigma} e^{-(x-\theta)^2/2\sigma^2} \left(\frac{dx}{dy}\right)^2 = 0$$

$$\frac{d^2x}{dy^2} - \frac{x-\theta}{\sigma^2}\left(\frac{dx}{dy}\right)^2 = 0 \qquad \text{at } y = \theta$$

$x = y$ is a solution of this equation because $dx/dy = 1$ and $d^2x/dy^2 = 0$. Hence $-(x-\theta)/\sigma^2 = 0$ at $x = \theta$. Thus X is a modal-unbiased estimate of θ.

Theorem 2

Let $x_1, x_2, \ldots, x_n$ be independent random variables. Let $t(X_1, \ldots, X_n)$ be a statistic having density function

$$f(T,\theta) = kh_1(T,\theta)h_2(T,\theta)$$

and $h_1(T,\theta) = \text{constant } [h_2(T,\theta)]^{-1}$ at $T = \theta$; furthermore

$$h_1''(T,\theta)h_2(T,\theta) + 2h_1'(T,\theta)h_2'(T,\theta) + h_1(T,\theta)h_2''(T,\theta) < 0$$

at $T = \theta$, where primes denote the order of differentiation with respect to T. Then t is a modal-unbiased estimate of θ.

Proof:

$$f(T,\theta) = kh_1(T,\theta)h_2(T,\theta)$$
$$f'(T,\theta) = kh_1'(T,\theta)h_2(T,\theta) + kh_1(T,\theta)h_2'(T,\theta)$$
$$h_1(T,\theta) = ch_2^{-1}(T,\theta) \qquad \text{at } T = \theta$$

Therefore

$$h_1'(T,\theta) = -\frac{ch_2'(T,\theta)}{h_2^2(T,\theta)} \qquad \text{at } T = \theta$$

Hence $f'(T,\theta) = 0$ at $T = \theta$ and $f''(T,\theta) < 0$ at $T = \theta$ by hypothesis. Hence t is a modal-unbiased estimate of θ.

Theorem 3

Let $X_1, X_2, \ldots, X_n$ be independent random variables with a common density function $f(x,\theta)$ and denote the density function of X_i by $f_i(x_i,\theta)$ and the product of the density function by L. The following are assumed:

(i) $(\partial \log L)/\partial \theta$ and $(\partial^2 \log L)/\partial \theta^2$ exist and are continuous for every θ in a range R, including the true value, and for almost all x. For every θ in R,

$$\left| \frac{\partial \log L}{\partial \theta} \right| < F_1(x), \qquad \left| \frac{\partial^2 \log L}{\partial \theta^2} \right| < F_2(x)$$

where F_1 and F_2 are integrable functions over $(-\infty, +\infty)$.

(ii) $(\partial^3 \log L)/\partial \theta^3$ exists, and $|(\partial^3 \log L)/\partial \theta^3| < M(x)$ with $E[M(x)] < k$, where k is a positive real number.

(iii) For every θ in R,

$$\int_{-\infty}^{\infty} -\frac{\partial^2 \log L}{\partial \theta^2} L \, dx = I(\theta)$$

is finite and nonzero, and $(\partial \log L)/\partial \theta = 0$ has a solution, say $\hat{\theta}$.

Then $\hat{\theta}$ is asymptotically a modal unbiased estimate.

Proof:

The asymptotic distribution of $\hat{\theta}$ is normal (see Chap. 9) with mean θ under the conditions of the theorem; thus the mode of the asymptotic distribution of $\hat{\theta}$ is θ. $\hat{\theta}$ is an asymptotically modal-unbiased estimate of θ. In fact, every best asymptotically normal (BAN) estimate is a modal-unbiased estimate.

7 □ RELATIVE EFFICIENCY OF THE MODAL - UNBIASED ESTIMATE AND MEAN - UNBIASED ESTIMATE

EXAMPLE 1

Let $X_1, X_2, \ldots, X_n$ be independent random variables with common density function

$$f(x,\theta) = \frac{1}{\theta} e^{-x/\theta} \qquad x > 0$$
$$\theta > 0$$

Then,

$$Y_1 = \frac{\sum_1^n X_i}{n}$$

is a mean-unbiased estimate of θ, and

$$Y_2 = \frac{\sum_1^n X_i}{n-1}$$

is a modal-unbiased estimate of θ. Let $L(\hat{\theta},\theta) = (1/\hat{\theta} - 1/\theta)^2$ be the loss function, where $\hat{\theta}$ is an estimate of θ. Then the expected loss of Y_2 is uniformly smaller than that of Y_1.

Proof:
It can easily be proved that

$$E\left(\frac{1}{Y_1} - \frac{1}{\theta}\right)^2 = \frac{n+2}{(n-1)(n-2)} \frac{1}{\theta^2} \quad \text{provided } n > 2$$

$$E\left(\frac{1}{Y_2} - \frac{1}{\theta}\right)^2 = \frac{1}{(n-2)\theta^2}$$

It is obvious that

$$E\left(\frac{1}{Y_1} - \frac{1}{\theta}\right)^2 > E\left(\frac{1}{Y_2} - \frac{1}{\theta}\right)^2 \quad \text{for every } \theta$$

Hence, the modal-unbiased estimate has uniformly smaller risk than the mean-unbiased estimate.

REMARK
This loss function was found to be appropriate for a practical industrial problem.

EXAMPLE 2
We have the Weibull density function with known $\alpha \geq 2$, namely,

$$f(x;\rho,\alpha) = \alpha\rho^\alpha x^{\alpha-1} e^{-(\rho x)^\alpha} \quad \begin{aligned} x &> 0 \\ \rho &> 0 \\ \alpha &\geq 2 \text{ but known} \end{aligned}$$

Then,

$$Y_1 = \frac{1}{\overline{(1 - 1/\alpha)X}}$$

is a mean-unbiased estimate of ρ,

$$Y_2 = \left(\frac{\alpha+1}{\alpha}\right)^{1/\alpha} \frac{1}{X}$$

is a modal-unbiased estimate of ρ. For $L(\hat{\rho},\rho) = (1/\hat{\rho} - 1/\rho)^2$ a loss function where $\hat{\rho}$ is an estimate of ρ,

$$E[L(Y_1,\rho)] > E[L(Y_2,\rho)] \quad \text{for every } \rho$$

Proof:
The result can easily be proved by simple calculations.

EXAMPLE 3
Let $X_1, X_2, \ldots, X_n$ be independent random variables with a common density function

$$f(x;\theta) = \frac{1}{\theta} e^{-x/\theta} \quad x > 0$$
$$\theta > 0$$

and let $L(\hat{\theta},\theta) = \hat{\theta}/\theta + \theta/\hat{\theta} - 2$ be a loss function, where $\hat{\theta}$ is an estimate of θ. $L(\hat{\theta},\theta)$ is a convex function in $\hat{\theta}$. Then the uniformly minimum-variance-unbiased estimate $Y_1 = \left(\sum_1^n X_i\right)/n$ and a modal-unbiased estimate $Y_2 = \left(\sum_1^n X_i\right)/n - 1$ have the same expected loss, even though $Y_1 < Y_2$ a. s.

Proof:
Let $Z = \sum_1^n X_i$. Then

$$E(Z) = n\theta \quad \text{and} \quad E(Z^{-1}) = \frac{1}{(n-1)\theta}$$

Therefore,

$$E[L(Y_1,\theta)] = 1 + \frac{n}{n-1} - 2 = \frac{1}{n-1}$$

and

$$E[L(Y_2,\theta)] = \frac{n}{n-1} + 1 - 2 = \frac{1}{n-1}$$

Thus

$$E[L(Y_1,\theta)] = E[L(Y_2,\theta)] = \frac{1}{n-1}$$

EXAMPLE 4

Let $X_1, X_2, X_3, \ldots, X_{2n+1}$ be independent random variables with negative-exponential density defined in Example 3. Let

$$X_{(1)} = \min_{1 \leq i \leq 2n+1} X_i$$

Then $X_{(1)}$ has a negative-exponential distribution, with density

$$f_{X_{(1)}}(x) = \begin{cases} \dfrac{2n+1}{\theta} \exp\left(-\dfrac{2n+1}{\theta} x\right) & 0 \leq x \leq \infty \\ 0 & \text{otherwise} \end{cases}$$

Hence, the median of the distribution of $X_{(1)}$ is $\theta(\log 2)/(2n+1)$. It follows that the median of the distribution of

$$Z_1 = \frac{2n+1}{\log 2} X_{(1)}$$

is θ. Thus, Z_1 is a median-unbiased estimate of θ, and

$$E(Z_1 - \theta)^2 = \theta^2 \frac{4n^2(2 - \log 2) + 8n + 1 + (n \log 2 - 1)^2}{n^2 \log^2 2}$$

It follows that $E_\theta(Y - \theta)^2 < E_\theta(Z_1 - \theta)^2$ for all $0 < \theta < \infty$ and all $n \geq 2$, where Y is the modal-unbiased estimate,

$$Y = \sum_{i=1}^{2n+1} \frac{X_i}{2n}$$

The estimator $Z = X_{(n+1)/\log 2}$ is *also* median-unbiased. In the same manner we can construct median-unbiased estimates and functions $Z_j = \lambda_j X_{(j)}, j = 1, \ldots, 2n+1$, of the order statistics $X_{(j)}$,

$$j = 1, \ldots, 2n+1.$$

One can continue and ask interesting questions concerning the minimum-risk linear combination of the order statistics, $\sum_{i=1}^{2n+1} \lambda_j X_{(j)}$, which yields a median-unbiased estimator. It should be mentioned that the minimum-risk linear combination which yields a mean-unbiased estimate of θ, for squared-error loss, is obviously the sample mean

$$\bar{X} = \frac{1}{2n+1} \sum_{j=1}^{2n+1} X_{(j)}$$

REMARK

For a negative exponential density function, the modal-unbiased estimate of θ is proportional to the uniformly minimum-variance mean-unbiased estimate and also to the maximum-likelihood estimate of θ; for the following distance criteria all three are equivalent:

(i) $J(\theta_1,\theta_2) = E_{\theta_1}\left[\log \dfrac{P(T,\theta_2)}{P(T,\theta_1)}\right] + E_{\theta_2}\left[\log \dfrac{P(T,\theta_1)}{P(T,\theta_2)}\right].$

(ii) $\rho(\theta_1,\theta_2) = \int_R \sqrt{P(T,\theta_1)P(T,\theta_2)}\, dT.$

The preceding discussion reveals that one should consider the concept of unbiased estimation from the point of view of decision theory.

PROBLEMS

1. Let X be a random variable with density function

$$P(X = k) = \frac{e^{-\lambda}\lambda^k}{k!(1-e^{-\lambda})} \qquad k = 1, 2, \ldots$$
$$\lambda > 0$$

Then prove that the only unbiased estimate of $1 - e^{-\lambda}$ is 0 for X an odd integer and 2 for X an even integer.

2. Let X be a random variable with density function

$$P(X = k) = \frac{e^{-\lambda}\lambda^k}{k!} \qquad k = 0, 1, 2, \ldots$$
$$\lambda > 0$$

Then prove that there does not exist an unbiased estimate of $1/\lambda$.

3. Let $X_1, X_2, \cdots, X_n$ be independent random variables with common density function

$$f_{\mu,\sigma^2}(x) = \frac{1}{\sqrt{2\pi\sigma^2}}\, e^{-(x-\mu)^2/2\sigma^2} \qquad -\infty < x < \infty$$
$$-\infty < \mu < \infty$$
$$\sigma > 0$$

Find the uniformly minimum-variance-unbiased estimate of the 80th percentile.

4. (Pitman) Let $X_1, X_2, \ldots, X_n$ be independent random variables with common density function $f(x - \theta)$, $-\infty < x < \infty$ and $-\infty < \theta < \infty$. State and prove precise conditions that

$$u(x_1, \ldots, x_n) = \frac{\int \theta f(x_1 - \theta) \cdots f(x_n - \theta)\, d\theta}{\int f(x_1 - \theta) \cdots f(x_n - \theta)\, d\theta}$$

has a minimum-variance-unbiased estimate among the class of all estimations $\phi(x_1, \ldots, x_n)$ with the translation property, i.e., estimators satisfying the condition

$$\phi(x_1 + a, \ldots, x_n + a) = \phi(x_1, \ldots, x_n) + a$$

where a is any real number.

5. Let X be a random variable with density function $f_\theta(x) = f(\theta - x)$. Let $t(\theta)$ be a function of parameter θ such that

(i) $t(\theta)$ belongs to $L^2(-\infty, \infty)$.
(ii) $f(\theta)$ belongs to $L(-\infty, \infty)$.
(iii) $T(u) = \dfrac{1}{\sqrt{2\pi}} \displaystyle\int_{-\infty}^{\infty} t(\theta) e^{i\theta u}\, d\theta$

and $\quad F(u) = \dfrac{1}{\sqrt{2\pi}} \displaystyle\int_{-\infty}^{\infty} f(\theta) e^{i\theta u}\, d\theta$

such that $T(u)/F(u)$ belongs to $L^2(-\infty, \infty)$.

Prove that

$$g(x) = \frac{1}{\pi} \int_{-\infty}^{\infty} \frac{T(u)}{F(u)} e^{-ixu}\, du$$

exists, is the unbiased estimate of $t(\theta)$, and belongs to $L^2(-\infty, \infty)$.

6. Let X be a random variable with density function $f(x, \theta) = f(x\theta)$, where $x > 0$ and $\theta > 0$. Let $t(\theta)$ be a function of θ such that:

(i) $t(\theta)$ belongs to $L^2(-\infty, \infty)$.
(ii) $f(\theta)$ belongs to $L(-\infty, \infty)$.
(iii) Let

$$L(S) = \int_0^\infty t(\theta) \theta^{S-1}\, d\theta \quad \text{and} \quad R(S) = \int_0^\infty f(\theta) \theta^{S-1}\, d\theta$$

Prove that

$$g(x) = \frac{1}{2\pi i} \int_{c-i\infty}^{c+i\infty} \frac{L(1-S)}{R(1-S)} x^{-S}\, dS$$

exists, belongs to $L^2(-\infty, \infty)$, and is the unbiased estimate of $t(\theta)$.

7. Let g_1 and g_2 be two unbiased estimates of a parameter θ with variances σ_1^2 and σ_2^2 and correlation ρ. What is the *best* unbiased linear combination of g_1 and g_2, and what is the variance of such a combination?

8. (Roy and Mitra) Let $X_1, X_2, \ldots, X_n$ be independent random variables with common density function

$$P(X = x) = \frac{a(x)\theta^x}{f(\theta)} \qquad x = 0, 1, 2, \ldots$$

$$\theta > 0$$
$$a(x) > 0$$
$$f(\theta) = \sum_0^\infty a(x)\theta^x$$
$$a(0) = 1$$

Let

$$C(t,n) = \sum \prod_{i=1}^n a(x_i)$$

Σ denoting summation over nonnegative integral values of $x_1, x_2, \ldots, x_n$ subject to $x_1 + x_2 + \cdots + x_n = t$. Prove that for any positive integer r,

$$U_r(t) = \begin{cases} 0 & \text{for } t < r \\ \dfrac{C(t-r,n)}{C(t,n)} & \text{for } t \geq r \end{cases}$$

and

$$t = \sum_1^n X_i$$

$U_r(T)$ is a uniformly minimum-variance-unbiased estimate of θ^r.

9. State and prove various methods of finding minimum-variance-unbiased estimates of a parameter.

10. (Kolmogorov) Let $X_1, X_2, \ldots, X_n$ be independent random variables, each having normal distribution with mean μ and variance σ^2, where μ and σ^2 are both unknown. Find the minimum-variance-unbiased estimate of proportion p, that is, the proportion p of the population which has values greater than some given value c. Recall that

$$\frac{1}{\sqrt{2\pi\sigma^2}} \int_c^\infty e^{-(x-\mu)^2/2\sigma^2}\, dx = p$$

11. Define density-unbiased estimate, illustrate by examples, and compare it with mean-, median-, and modal-unbiased estimates by taking various loss functions.

REFERENCES

1 BARANKIN, E. W.: Locally Best Unbiased Estimates, *Ann. Math. Statist.*, vol. 20, pp. 477–501, 1949.
2 BIRNBAUM, A.: On the Foundation of Statistical Inference, *J. Am. Statist. Assoc.*, vol. 57, pp. 269–326, 1962.
3 BLACKWELL, D.: Conditional Expectation and Unbiased Sequential Estimation, *Ann. Math. Statist.*, vol. 18, pp. 105–110, 1947.
4 BLACKWELL, D., and M. A. GIRSCHICK: "Theory of Games and Statistical Decision Functions," John Wiley & Sons, Inc., New York, 1954.
5 CRAMÉR, H.: "Mathematical Methods of Statistics," Princeton University Press, Princeton, N.J., 1946.
6 FRASER, D. A. S.: "Nonparametric Methods in Statistics," John Wiley & Sons, Inc., New York, 1957.
7 GIRSCHICK, M. A., F. MOSTELLER, and L. J. SAVAGE: Unbiased Estimate for Certain Binomial Sampling Problems with Applications, *Ann. Math. Statist.*, vol. 17, pp. 13–23, 1946.
8 KOLMOGOROV, A. N.: Unbiased Estimates (in Russian), *Izv. Akad. Nauk USSR*, vol. 14, p. 303, 1950.
9 LEHMANN, E. L., and H. SCHEFFÉ: Completeness, Similar Regions and Unbiased Estimation, *Sankhyā, ser. A*, vol. 10, pp. 305–340, 1950.
10 LEHMANN, E. L.: "Notes on Theory of Estimation," University of California Press, Berkeley, Calif., 1950.
11 PATIL, G. P.: Minimum Variance Unbiased Estimation and Certain Problems of Additive Number Theory, *Ann. Math. Statist.*, vol. 34, pp. 1050–1056, 1963.
12 RAO, C. R.: Information and Accuracy Attainable in Estimation of Statistical Parameters, *Bull. Calif. Math. Soc.*, vol. 37, pp. 81–91, 1945.
13 RAO, C. R.: Sufficient Statistics and Minimum Variance Estimates, *Proc. Cambridge Phil. Soc.*, vol. 45, pp. 213–218, 1949.
14 ROY, T., and S. MITRA: Unbiased Minimum Variance Estimation in a Class of Discrete Distributions, *Sankhyā, ser. A*, vol. 18, pp. 371–378, 1957.
15 TATE, R. F.: Unbiased Estimation; Functions of Estimation and Scale Parameters, *Ann. Math. Statist.*, vol. 30, pp. 341–366, 1959.
16 WASAN, M. T.: Theory of Modal Unbiased Estimation, abstract, *Ann. Math. Statist.*, vol. 36, p. 1324, 1965.
17 WASAN, M. T.: Sequential Optimum Procedures for Unbiased Estimation of a Binomial Parameter, *Technometrics*, vol. 6, pp. 259–272, 1964.
18 WALD, A.: "Statistical Decision Functions," John Wiley & Sons, Inc., New York, 1950.

chapter Eight

A Lower Bound of the Variance of an Estimate

1 □ INTRODUCTION

In the previous chapter we discussed various methods of obtaining a minimum-variance-unbiased estimate. In this chapter we are going to discuss how one can find a lower bound of the variance of an estimate. The main result is known in statistical literature as the *Cramér-Rao inequality*. We shall follow Wolfowitz [12] because his conditions of regularity are easy to check. For the case when a density function depends on more than one parameter we discuss a method of obtaining a lower bound.

The modification of the Cramér-Rao inequality due to Chapman, Kiefer, and Robbins is also given and illustrated by an example, and we discuss Bhattacharyya bounds of the variance of an estimate. Many other interesting results are available, e.g., the result of Fraser and Guttman [7] and of Fend [6], but we shall not discuss them here. These methods of finding a lower bound of the variance are also useful in determining optimum stopping rules for sequential-estimation procedures.

2 □ THE CRAMÉR - RAO INEQUALITY

We state and prove the Cramér-Rao inequality for finding a lower bound of the variance of the estimate.

Wolfowitz's regularity conditions

Let $X_1, \ldots, X_n$ be independent random variables each having the same density $p_\theta(x)$ absolutely continuous with respect to a σ-additive measure μ. An estimate $\delta(X_1, \ldots, X_n)$, not necessarily unbiased, is wanted for the real parameter θ.

(i) θ lies in an open interval Ω of the real line; Ω may be infinite.
(ii) $\partial p_\theta(x)/\partial \theta$ exists for all $\theta \in \Omega$.
(iii) $\int p_\theta(x_1) p_\theta(x_2), \ldots, p_\theta(x_n)\, d\mu(x_1), \ldots, d\mu(x_n)$ can be differentiated under the integral sign.
(iv) $E_\theta \left[\dfrac{\partial \log p_\theta(X)}{\partial \theta} \right]^2 > 0 \quad$ for every $\theta \in \Omega$.
(v) $\int \delta(x_1, \ldots, x_n) p_\theta(x_1) p_\theta(x_2), \ldots, p_\theta(x_n)\, d\mu(x_1), \ldots, d\mu(x_n)$ can be differentiated under the integral sign.

Theorem

If $E_\theta[\delta(X_1, \ldots, X_n)] = \theta + b(\theta)$ and regularity conditions (i) to (v) are satisfied, then

$$\sigma_\delta^2(\theta) \geq \frac{[1 + b'(\theta)]^2}{nE_\theta[\partial/\partial\theta \log p_\theta(X)]^2}$$

Proof:

$$\int \delta(x_1, \ldots, x_n) p_\theta(x_1), \ldots, p_\theta(x_n)\, d\mu(x_1), \ldots, d\mu(x_n) = \theta + b(\theta)$$

Differentiating, we obtain, using condition (v),

$$\int \delta(x_1, \ldots, x_n) \sum_{i=1}^{n} \left[\frac{1}{p_\theta(x_i)} \frac{\partial}{\partial \theta} p_\theta(x_i) \right] p_\theta(x_1), \ldots, p_\theta(x_n)\, d\mu(x_1),$$

$$\ldots, d\mu(x_n) = E_\theta \left[\delta(X_1, \ldots, X_n) \sum_{i=1}^{n} \frac{\partial \log p_\theta(X_i)}{\partial \theta} \right] = 1 + b'(\theta)$$

Similarly, differentiating the equation $E_\theta(1) = 1$ and using condition (iii), we have

$$E_\theta\left[\sum_{i=1}^{n} \frac{\partial}{\partial\theta} \log p_\theta(X_i)\right] = 0$$

Writing T for $\sum_{i=1}^{n} \frac{\partial}{\partial\theta} \log p_\theta(X_i)$, we have so far shown that

$$E_\theta(\delta T) = 1 + b'(\theta)$$
$$E_\theta(T) = 0$$

We have, therefore, for the covariance of δ and T

$$\text{cov}(\delta, T) = 1 + b'(\theta)$$

But since the square of a correlation coefficient cannot exceed 1, we have

$$[\text{cov}(\delta, T)]^2 \leq \sigma_\delta^2 \sigma_T^2$$
$$[1 + b'(\theta)]^2 \leq \sigma_\delta^2 n E\left[\frac{\partial}{\partial\theta} \log p_\theta(X)\right]^2$$

Hence

$$\sigma_\delta^2 \geq \frac{[1 + b'(\theta)]^2}{nE[\partial/\partial\theta \log p_\theta(X)]^2}$$

Clearly this can become an equality only if $\rho_{\delta,T}^2 = 1$ where $\rho_{\delta,T}$ is correlation between δ and T, that is, only if δ is a linear function of T for almost all x.

We shall give some illustrative examples.

REMARK

Generally it is easy to check regularity conditions (i), (ii), and (iv). To check conditions (iii) and (v) one can use the following result of Cramér [5]. Let us assume that (i), (ii), and (iv) are true. Suppose that for each $\theta_0 \in \Omega$, there are numbers a and b and μ-integrable functions G_1 and G_2 such that $\theta_0 \in (a,b) \subset \Omega$ and such that

$$\left|\frac{\partial p_\theta(x)}{\partial\theta}\right| \leq G_1(x) \quad \theta \in (a,b)$$
$$x \in W - N$$

for some N such that
$$\mu(N) = 0$$
$$\left|\delta(x) \frac{\partial p_\theta(x)}{\partial \theta}\right| \leq G_2(x) \qquad \theta \in (a,b)$$
$$x \in W - N$$

then (iii) and (v) are satisfied provided that $\int_W G_i(x) \, d\mu(x) < \infty$, $i = 1, 2$.
We now use this result in the following illustrative example.

EXAMPLE 1
Let X be normally distributed with mean θ and variance 1, where $\theta \in \Omega = (-\infty, \infty)$. μ is the Lebesgue measure. Thus
$$p_\theta(x) = \frac{1}{\sqrt{2\pi}} e^{-(x-\theta)^2/2} = \frac{1}{\sqrt{2\pi}} e^{-\theta^2/2} e^{-x^2/2} e^{\theta x}$$
$$\frac{\partial p_\theta(x)}{\partial \theta} = (x - \theta) p_\theta(x)$$
$$\frac{\partial \log p_\theta(x)}{\partial \theta} = x - \theta$$

Then
$$\int_{-\infty}^{\infty} \frac{\partial p_\theta(x)}{\partial \theta} \, dx = e^{-\theta^2/2} \int_{-\infty}^{\infty} h(x) e^{\theta x} \, dx$$

where
$$h(x) = (x - \theta) e^{-x^2/2}$$

for a given value of θ.
$$\int_{-\infty}^{\infty} \delta(x) \frac{\partial p_\theta(x)}{\partial \theta} \, dx = e^{-\theta^2/2} \int_{-\infty}^{\infty} \delta(x) h(x) e^{\theta x} \, dx$$

The desired result holds if for each h, the following is satisfied
$$\int_{-\infty}^{\infty} |h(x)| e^{\theta x} \, dx < \infty \qquad \text{for } \theta \in \Omega$$

Let $\theta_0 \in \Omega$ be a real number, $\theta_0 \in (-b,b)$, and $x \in W - N$. Then
$$G(x) = (e^x + e^{-x}) |h(x)| (e^{bx} + e^{-bx})$$
$$\int_{-\infty}^{\infty} G(x) \, dx = \int_{-\infty}^{\infty} |h(x)| e^{(b+1)x} \, dx + \int_{-\infty}^{\infty} |h(x)| e^{-(b+1)x} \, dx$$
$$+ \int |h(x)| e^{(b-1)x} \, dx + \int_{-\infty}^{\infty} |h(x)| e^{(-b+1)x} \, dx < \infty$$
$$\left| h(x) \frac{\partial e^{\theta x}}{\partial \theta} \right| = |x| |h(x)| e^{\theta x} \leq G(x) \qquad \text{for all } x$$
$$\text{all } \theta \in (-b,b)$$

Thus we can differentiate the integral under the integral sign. Then let $\delta(x) = X$, $E(X) = \theta$,

$$\sigma_\delta^2 \geq \frac{1}{E_\theta\{[\partial \log p_\theta(X)]/\partial \theta\}^2} = \frac{1}{E_\theta(X - \theta)^2} = 1$$

and var $(X) = 1$. Hence, it attains the Cramér-Rao lower bound of variance.

EXAMPLE 2
Let X have density function

$$p_\theta(x) = \binom{n}{x} \theta^x (1 - \theta)^{n-x}$$

where n is a known positive integer and $x = 0, 1, 2, \ldots, n$ and $\theta \in \Omega = (0,1)$. Conditions (i), (ii), and (iv) can easily be checked to be satisfied. Let us consider (v).

$$f(\theta) = \sum_{x=0}^{n} \delta(x) p_\theta(x) = \sum_{x=0}^{n} \delta(x) \binom{n}{x} \theta^x (1 - \theta)^{n-x}$$

Since $f(\theta)$ is a polynomial in θ, it can be differentiated. Similarly, one can conclude that (iii) holds true. Let $\delta(x) = x/n$; then

$$\sigma_\delta^2 \geq \frac{1}{E_\theta\left\{\dfrac{[\partial \log p_\theta(X)]}{\partial \theta}\right\}^2} = \frac{1}{n} \theta(1 - \theta)$$

Since $\sigma_\delta^2 = (1/n)\theta(1 - \theta)$, $\delta(X) = X/n$ attains the Cramér-Rao lower bound of variance.

EXAMPLE 3
Let X have the density function

$$p_\theta(x) = \begin{cases} \dfrac{1}{\theta} & \text{if } 0 \leq x \leq \theta \\ 0 & \text{otherwise} \end{cases} \quad \text{where } \theta \in \Omega = (0, \infty)$$

In this case, $\partial p_\theta/\partial \theta$ exists if $\theta \neq x$ but does not exist if $\theta = x$; thus condition (ii) is not satisfied. Hence, we cannot use the established result. We shall show later how one can handle this situation.

3 □ CRAMÉR - RAO LOWER BOUND WITH NUISANCE PARAMETERS

In the previous section, the Cramér-Rao bound was developed for density functions depending on one parameter. Obviously, its use is severely limited; hence, in this section, it will be extended to the case in which more than one parameter is present. However, in order to extend the Cramér-Rao result, the following result, due to Lehmann and Hodges, must first be proved.

Theorem 1

Let $r_1, r_2, \ldots, r_k$ and $t_1, t_2', \ldots, t_l$ be two sets of random variables, each being linearly independent. Let the covariance matrix of the r's be $P = [p_{ij}]$, of the t's be $T = [\tau_{ij}]$ and of the r's with the t's be $\Gamma = [\gamma_{ij}]$. Then $T - \Gamma'P\Gamma$ is positive semidefinite.

Proof:

It must be shown that for any vector **L**

$$\mathbf{L}(T - \Gamma'P^{-1}\Gamma)L' \geq 0$$

In order to simplify the proof and without loss of generality, it is assumed that $E(r_m) = E(t_n) = 0$ for all admissible m and n. Using the familiar technique of applying Schwartz' inequality to the definition of covariance, it may be said that

$$\text{cov}^2 \left(\sum_{m=1}^{k} u_m r_m, \sum_{n=1}^{l} l_n t_n \right) \leq \text{var} \left(\sum_{m=1}^{k} u_m r_m \right) \text{var} \left(\sum_{n=1}^{l} l_n t_n \right)$$

where u_m are the elements of some vector **U**. Performing the indicated operations in the above equation,

$$\left[E \left(\sum_{m=1}^{k} \sum_{n=1}^{l} u_m l_n r_m t_n \right) \right]^2 \leq E \left(\sum_{m=1}^{k} \sum_{n=1}^{k} u_m u_n r_m r_n \right) E \left(\sum_{m=1}^{l} \sum_{n=1}^{l} l_m l_n t_m t_n \right)$$

$$\left(\sum_{m=1}^{k} \sum_{n=1}^{l} u_m l_n \gamma_{mn} \right)^2 \leq \left(\sum_{m=1}^{k} \sum_{n=1}^{k} u_m u_n p_{mn} \right) \left(\sum_{m=1}^{l} \sum_{n=1}^{l} l_m l_n \tau_{mn} \right)$$

or, in matrix form,

$$(\mathbf{U}\Gamma L')^2 \leq (\mathbf{U}P\mathbf{U}')(\mathbf{L}T\mathbf{L}')$$

Now setting $\mathbf{U} = \mathbf{L}\Gamma' P^{-1}$

$$(\mathbf{L}\Gamma' P^{-1} T \mathbf{L}')^2 \leq (\mathbf{L}\Gamma' P^{-1} T \mathbf{L}')(\mathbf{L} T \mathbf{L}')$$

Since the r's are linearly independent, the P matrix is positive definite, and from this it follows that $\mathbf{L}\Gamma P^{-1} T \mathbf{L}'$ is also positive definite. Therefore

$$\mathbf{L}\Gamma' P^{-1} T \mathbf{L}' \leq \mathbf{L} T \mathbf{L}'$$

Because T is positive semidefinite it follows that for all L

$$\mathbf{L}(T - \Gamma' P^{-1} \Gamma)\mathbf{L}' \geq 0$$

EXAMPLE 1

Consider the case of $l = 1$, $\mathbf{L} = \mathbf{L}'$, T and $\Gamma' P \Gamma$ are real numbers, and the results show that

$$T = \sigma_t^2 \geq \Gamma' P^{-1} \Gamma$$

Using the previous theorem, it is now possible to extend the Cramér-Rao result to the case when nuisance parameters are present.

Theorem 2

Let $r_1, r_2, \ldots, r_k$ be linearly independent random variables, each with mean zero and covariance matrix $P = [p_{mn}]$. If t is some random variable such that $E(r_1 t) = 1$ and $E(r_m t) = 0$, $m = 2, 3, \ldots, k$, then

$$\sigma_t^2 \geq \frac{\text{cofactor of } p_{11}}{|p_{mn}|}$$

Proof:
$\Gamma' = (1, 0, \ldots, 0)$ and $\Gamma' P^{-1} \Gamma = (1, 0, \ldots, 0) P^{-1} (1, 0, \ldots, 0)'$ which is the leading element of P^{-1}, and the result follows from the previous theorem.

EXAMPLE 2

Given a sample of n independent random variables, each with common density function $p_{\theta_1, \ldots, \theta_k}(x)$, find a lower bound on the variance of the unbiased estimate of θ_1. (Assume that the regularity conditions of Wolfowitz are satisfied.)

$$\int \prod_{\alpha=1}^{n} p_{\theta_1, \ldots, \theta_k}(x_\alpha) dm(x_\alpha) = 1$$

Differentiating with respect to θ_i gives

$$\int \left[\sum_{\alpha=1}^{n} \frac{\partial}{\partial \theta_i} \log p_{\theta_1,\ldots,\theta_k}(x_\alpha)\right] \prod_{\alpha=1}^{n} p_{\theta_1,\ldots,\theta_k}(x_\alpha) dm(x_\alpha) = 0$$

or if

$$r_i = \sum_{\alpha=1}^{n} \frac{\partial}{\partial \theta_i} \log p_{\theta_1,\ldots,\theta_k}(x_\alpha)$$

$$E(r_i) = 0$$

Let $\delta(x_1, x_2, \ldots, x_n) = \Delta$ be the estimate on which the bound is desired; then

$$\int \Delta \prod_{\alpha=1}^{n} p_{\theta_1,\ldots,\theta_k}(x_\alpha) \, dm(x_\alpha) = \theta_1$$

and differentiating with respect to θ_i gives

$$\int \Delta \sum_{\alpha=1}^{n} \frac{\partial}{\partial \theta_i} \log p_{\theta_1,\ldots,\theta_k}(x_\alpha) \prod_{\alpha=1}^{n} p_{\theta_1,\ldots,\theta_k}(x_\alpha) \, dm(x_\alpha)$$

$$= \begin{cases} 1 & i = 1 \\ 0 & i = 2, 3, \ldots, k \end{cases}$$

or

$$E(r_1 \Delta) = 1 \quad \text{and} \quad E(r_i \Delta) = 0 \quad \text{for } i = 2, 3, \ldots, k$$

Let us assume that

$$P_{mk} = E(r_m r_l) = n\alpha_{ml}$$

$$= nE\left[\frac{\partial}{\partial \theta_m} \log p_{\theta_1,\ldots,\theta_k}(X) \frac{\partial}{\partial \theta_l} \log p_{\theta_1,\ldots,\theta_k}(X)\right]$$

The theorem gives the desired result

$$\sigma_\delta^2(\theta_1, \ldots, \theta_k) \geq \frac{\text{cofactor of } \alpha_{11}}{n|\alpha_{ij}|}$$

EXAMPLE 3

Let $X_1, X_2, \ldots, X_n$ be independent random variables with density function

$$p_{\theta_1,\theta_2}(x) = \frac{1}{\sqrt{2\pi\theta_2}} e^{-1/2\theta_2 (x-\theta_1)^2} \quad -\infty < x < \infty$$

$$-\infty < \theta_1 < \infty$$
$$\theta_2 > 0$$

We find a lower bound for the variance of unbiased estimate of θ_1. One can easily compute

$$E\left[\frac{\partial^2 \log p_{\theta_1,\theta_2}(X)}{\partial \theta_1{}^2}\right] = -\frac{1}{\theta_2}$$

$$E\left[\frac{\partial^2 \log p_{\theta_1,\theta_2}(X)}{\partial \theta_1\, \partial \theta_2}\right] = 0$$

$$E\left[\frac{\partial^2 \log p_{\theta_1,\theta_2}(X)}{\partial^2 \theta_2}\right] = -\frac{1}{2\theta_2{}^2}$$

Then

$$\delta_\delta{}^2(\theta_1,\theta_2) \geq \frac{\left|-\dfrac{1}{2\theta_2{}^2}\right|}{n\left|\begin{array}{cc} -\dfrac{1}{\theta_2} & 0 \\ 0 & -\dfrac{1}{2\theta_2{}^2} \end{array}\right|} = \frac{\theta_2}{n}$$

For the function, we cannot establish a lower bound of the variance. The following modification is used to remedy the situation.

4 □ CHAPMAN - ROBBINS - KIEFER INEQUALITY

Let $\{W, \mathcal{C}, (p_\theta, \theta \in \Omega)\}$ be the sample space and μ be a Lebesgue measure. Assume that if $\theta \neq \theta^*$, then p_θ is different from p_{θ^*} in the sense that

$$\mu\{x | p_{\theta^*}(x) \neq p_\theta(x)\} > 0$$

Let $S(\theta)$ be the set of all x's such that $p_\theta(x) > 0$.

$$S_{(\theta)} \subset W \quad \text{and} \quad \int_W h(x) p_\theta(x)\, d\mu(x) = \int_{S(\theta)} h(x) p_\theta(x)\, d\mu(x)$$

Theorem

Suppose that if $\theta \in \Omega$, there is a $\theta^* \in \Omega$ such that $\theta^* \neq \theta$ and $S(\theta^*) \subset S(\theta)$. If δ is an unbiased estimate of γ, then

$$\operatorname{var}_\theta \delta \geq \sup_{\{\theta^* | \theta^* \neq \theta,\, s(\theta^*) \subset S(\theta)\}} \frac{[\gamma(\theta^*) - \gamma(\theta)]^2}{E_\theta[(p_{\theta^*} - p_\theta)/p_\theta]^2}$$

for each $\theta \in \Omega$.

Proof:

$$\theta \in \Omega$$

$$g_\theta(x) = \begin{cases} \dfrac{p_{\theta^*}(x) - p_\theta(x)}{p_\theta(x)} & \text{if } p_\theta(x) > 0 \\ 0 & \text{otherwise} \end{cases}$$

$$C(\theta) = \gamma(\theta^*) - \gamma(\theta) \qquad \theta^* \text{ is some element such that } \theta^* \neq \theta$$
$$S(\theta^*) \subset S(\theta)$$

$$E_\theta[g_\theta(x)] = \int_{s(\theta)} \frac{p_{\theta^*}(x) - p_\theta(x)}{p_\theta(x)} p_\theta(x) \, d\mu(x)$$
$$= 0$$

$$\text{cov}(\delta, g) = E\left[(\delta - \gamma) \frac{p_{\theta^*} - p_\theta}{p_\theta} \right]$$
$$= E\left[\frac{\delta(p_{\theta^*} - p_\theta)}{p_\theta} \right]$$
$$= E\left(\frac{\delta p_{\theta^*}}{p_\theta} \right) - E\left(\frac{\delta p_\theta}{p_\theta} \right)$$
$$= \gamma(\theta^*) - \gamma(\theta)$$

Therefore

$$\text{var } \delta \geq \sup_{\{\theta^* | \theta^* \neq \theta, S(\theta^*) \subset S(\theta)\}} \frac{[\gamma(\theta^*) - \gamma(\theta)]^2}{E[(p_{\theta^*} - p_\theta)/p_\theta]^2}$$

EXAMPLE

X is a random variable with distribution

$$p_\theta(x) = \begin{cases} \dfrac{1}{\theta} & \text{if } 0 \leq x \leq \theta \\ 0 & \text{otherwise} \end{cases}$$

Let $\gamma(\theta) = \theta$

$$E\left(\frac{p_{\theta^*}}{p_\theta} \right)^2 = \int_0^\theta \left[\frac{p_{\theta^*}(x)}{p_\theta(x)} \right]^2 \frac{1}{\theta} \, dx$$

If $S(\theta^*) \subset S(\theta)$,

$$\int_0^{\theta^*} \frac{\theta}{\theta^{*2}} \, dx = \frac{\theta}{\theta^*}$$

$$\text{var}_\theta \, \delta \geq \sup_{\{\theta^* | \theta^* \neq \theta, S(\theta^*) \subset S(\theta)\}} \frac{(\theta^* - \theta)^2}{\theta/\theta^* - 1}$$

$$= \sup_{\{\theta^* | \theta^* \neq \theta, S(\theta^*) \subset S(\theta)\}} \frac{\theta^*(\theta^* - \theta)^2}{\theta - \theta^*}$$

$$= \sup_{\{\theta^* | \theta^* \neq \theta, S(\theta^*) \subset S(\theta)\}} \theta^*(\theta - \theta^*) = \frac{\theta^2}{4}$$

$$\text{var}(2X) = \frac{\theta^2}{3} > \frac{\theta^2}{4}$$

144 PARAMETRIC ESTIMATION

It can be shown that $2X$ is a uniformly minimum-variance-unbiased estimate.

In the next section we discuss a method of finding a lower bound of the variance of an estimate given by Bhattacharyya [3], which, in a way, is a generalization of the method of Cramér-Rao inequality.

5 □ BHATTACHARYYA BOUNDS

Let $X_1, \ldots, X_n$ be independent random variables with common density function $f_\theta(x)$. We would like to determine a lower bound for the variance of unbiased estimate of θ. Let $U = \delta(X_1, \ldots, X_n)$ be any such estimate. We assume all conditions of the Cramér-Rao inequality. Thus we have

$$\int_\Omega \frac{1}{\prod_1^n f_\theta(x_i)} \frac{\partial^i}{\partial \theta^i} \left[\prod_1^n f_\theta(x_i) \right] \prod_1^n f_\theta(x_i) \prod_1^n dx_i = 0$$

Let

$$S_i = \frac{1}{\prod_1^n f_\theta(x_i)} \frac{\partial^i}{\partial \theta^i} \left[\prod_1^n f_\theta(x_i) \right]$$

Since

$$\int_\Omega \delta(x_1, \ldots, x_n) \prod_1^n f_\theta(x_i) \prod_1^n dx_i = \theta$$

gives us

$$E(S_1\delta) = 1, \quad E(S_i\delta) = 0 \quad \text{for } i = 2, \ldots, k$$

and $v_{ij} = E(S_i S_j)$, we have

$$\text{var}_\delta \geq L_k = \frac{\begin{vmatrix} v_{22} & \cdots & v_{2k} \\ \cdots & \cdots & \cdots \\ v_{k2} & \cdots & v_{kk} \end{vmatrix}}{\begin{vmatrix} v_{11} & \cdots & v_{1k} \\ \cdots & \cdots & \cdots \\ v_{k1} & \cdots & v_{kk} \end{vmatrix}}$$

Proof:

$$\begin{vmatrix} \text{var } \delta & E(S_1\delta) & E(S_2\delta) & \cdots & E(S_k\delta) \\ E(S_1\delta) & E(S_1S_1) & E(S_2S_1) & \cdots & E(S_kS_1) \\ \cdots & \cdots & \cdots & \cdots & \cdots \\ E(S_k\delta) & \cdots & \cdots & \cdots & E(S_kS_k) \end{vmatrix} \geq 0$$

Hence

$$\begin{vmatrix} \text{var}(\delta) & 1 & 0 & \cdots & 0 \\ 1 & v_{11} & v_{12} & \cdots & v_{1k} \\ 0 & v_{21} & v_{22} & \cdots & v_{2k} \\ \cdots & \cdots & \cdots & \cdots & \cdots \\ 0 & v_{k1} & v_{k2} & \cdots & v_{kk} \end{vmatrix}$$

$$= \text{var } \delta \begin{vmatrix} v_{11} & v_{12} & \cdots & v_{1k} \\ \cdots & \cdots & \cdots & \cdots \\ v_{k1} & v_{k2} & \cdots & v_{kk} \end{vmatrix} - \begin{vmatrix} v_{22} & \cdots & v_{2k} \\ \cdots & \cdots & \cdots \\ v_{k2} & \cdots & v_{kk} \end{vmatrix} \geq 0$$

$$\text{var } \delta \geq \frac{\begin{vmatrix} v_{22} & \cdots & v_{2k} \\ \cdots & \cdots & \cdots \\ v_{2k} & \cdots & v_{kk} \end{vmatrix}}{\begin{vmatrix} v_{11} & \cdots & v_{1k} \\ \cdots & \cdots & \cdots \\ v_{k1} & \cdots & v_{kk} \end{vmatrix}}$$

EXAMPLE

$$f_\theta(x) = \theta(1-\theta)^x \qquad x = 0, 1, 2, \ldots$$

$$\frac{\partial f_\theta(x)}{\partial \theta} = \frac{1}{\theta} f(x) - x f(x-1)$$

$$\frac{\partial^2 f_\theta(x)}{\partial \theta^2} = \frac{2x}{\theta} f(x-1) + x(x-1) f(x-2)$$

$$\frac{\partial^3 f_\theta(x)}{\partial \theta^3} = \frac{3x(x-1)}{\theta} f(x-2) - x(x-1)(x-2) f(x-3)$$

$$S_1 = \frac{1}{f(x)} \frac{\partial f}{\partial \theta} = -\frac{1}{1-\theta}\left(x - \frac{1-\theta}{\theta}\right)$$

$$S_2 = \frac{1}{f(x)} \frac{\partial^2 f}{\partial \theta^2} = \frac{x^2}{(1-\theta)^2} - \frac{[1+(1-\theta)]x}{\theta(1-\theta)^2}$$

$$E(X) = \frac{1-\theta}{\theta}$$

$$E(X^2) = \frac{(1-\theta) + (1-\theta)^2}{\theta^2}$$

$$E(X^3) = \frac{(1-\theta) + 4(1-\theta)^2 + (1-\theta)^3}{\theta^3}$$

$$E(X^4) = \frac{(1-\theta) + 11(1-\theta)^2 + 11(1-\theta)^3 + (1-\theta)^4}{\theta^4}$$

$$E(S_1^2) = \frac{1}{\theta^2(1-\theta)}$$

$$E(S_1 S_2) = -\frac{2}{\theta^3(1-\theta)}$$

$$E(S_2^2) = \frac{4[1+(1-\theta)]}{\theta^4(1-\theta)^2}$$

$$L_1 = \frac{1}{E(S_1^2)} = \theta^2(1-\theta)$$

which is less than $\theta(1-\theta)$.

$$v = (v_{ij}) = \begin{vmatrix} \dfrac{1}{\theta^2(1-\theta)} & \dfrac{-2}{\theta^3(1-\theta)} \\ \dfrac{-2}{\theta^3(1-\theta)} & \dfrac{4[1+(1-\theta)]}{\theta^4(1-\theta)^2} \end{vmatrix}$$

$$|v| = \frac{4}{\theta^6(1-\theta)^3}$$

$$L_2 = \frac{4[1+(1-\theta)]}{\theta^4(1-\theta)^2} \frac{\theta^6(1-\theta)^3}{4} = \theta^2(1-\theta)[1+(1-\theta)]$$

PROBLEMS

1. Let $X_1, X_2, \ldots, X_n$ be independent random variables each having common Poisson density function

$$P_\theta(x) = e^{-\theta}\frac{\theta^x}{x!} \qquad \begin{matrix} x = 0, 1, 2, \ldots \\ \theta \in \Omega = (0, \infty) \end{matrix}$$

Let $\delta(X_1, \ldots, X_n)$ be an unbiased estimate of θ; then check the regularity conditions for the Cramér-Rao inequality and find a lower bound to the variance of δ.

2. Let $x_1, x_2, \ldots, x_n$ be independent random variables with common density function

$$P(x_i = 1) = \theta \quad \begin{array}{l} 0 < \theta < 1 \\ i = 1, \ldots, n \end{array}$$
$$P(x_i = 0) = 1 - \theta$$

Let $\delta(x_1, \ldots, x_n)$ be an unbiased estimate of θ; then find the Cramér-Rao lower bound to the variance of δ and check the regularity conditions.

3. Let x be normally distributed with mean 0 and unknown variance σ^2, and let $\delta(x)$ be an unbiased estimate of σ^2; then find the Cramér-Rao lower bound of the variance of $\delta(x)$ and check the regularity conditions.

4. Let $X_1, X_2, \ldots, X_n$ be independent random variables with density function

$$P_{\theta_1, \theta_2}(x) = \frac{1}{\sqrt{2\pi\theta_2}} e^{-(1/2\theta_2)(x-\theta_1)^2} \quad \begin{array}{l} -\infty < x < \infty \\ -\infty < \theta_1 < \infty \\ 0 < \theta_2 < \infty \end{array}$$

Let $\delta(x_1, \ldots, x_n)$ be an unbiased estimate of θ_2; then find the Cramér-Rao lower bound to its variance.

5. Let $x_1, x_2, \ldots, x_n$ be independent random variables with common density function

$$P_\theta(x) = \begin{cases} e^{-(x-\theta)} & x \geq \theta > 0 \\ 0 & \text{otherwise} \end{cases}$$

Let δ be an unbiased estimate of θ, and find the Chapman-Robbins-Kiefer lower bound of variance of δ.

6. Let $x_1, x_2, \ldots, x_n$ be independent random variables normally distributed with mean zero and unknown standard derivation σ, and let $\delta(x_1, \ldots, x_n)$ be an unbiased estimate of σ; then find the Cramér-Rao lower bound of variance δ.

7. Let[1] x be a random variable with continuous density function $P_\theta(x)$ such that the function is specified when θ is specified. Let θ be a random variable with distribution $\lambda(\theta)$ defined over a nondegenerate range Ω.

---

[1] For Probs. 7 to 9 refer to Gart [8].

148 PARAMETRIC ESTIMATION

Let $\delta = \delta(x)$ be an estimate of θ, functionally independent of θ. Let $\psi(\theta) = E(\delta|\theta)$.

(i) $\partial P_\theta(x)/\partial \theta$ exists for all $\theta \in \Omega$.
(ii) $|\partial P_\theta(x)/\partial \theta| < G_1(x)$.
(iii) G_1 and δG are integrable over the entire line R_1, the range of x, which is independent of θ.

Then prove that

(iv) $\operatorname{var}(\delta - \theta) EE \left\{ \left[\dfrac{\partial \log P_\theta(x)}{\partial \theta} \right]^2 \Big| \theta \right\} \geq E^2 \left[\dfrac{\partial \psi(\theta)}{\partial \theta} \right]$

where EE stands for the expectation with respect to x for fixed θ and then with respect to θ.

8. State[1] and prove Prob. 7 when the density function is discrete.

9. Assume[1] the setup of Prob. 7 and let

(i) $P_r\{E[(\delta - \theta)^2|\theta] = c_1\} = 1$.
(ii) $P_r[\psi(\theta) = c_2] = 1$.
(iii) $P_r(\theta = c_3) = 1$.
(iv) $P_r\left[\dfrac{\partial \psi(\theta)}{\partial \theta} = c_4 \right] = 1$.
(v) $P_r\left(EE \left\{ \left[\dfrac{\partial \log P_\theta(x)}{\partial \theta} \right]^2 \Big| \theta \right\} = c_5 \right) = 1$

when c_i, $i = 1, 2, \ldots, 5$ are constants.

Then prove that the equality in (iv) holds if and only if δ is a sufficient statistic for θ.

REFERENCES

1 AITKEN, A. C., and H. SILVERSTONE: On the Estimation of Statistical Parameters, *Proc. Roy. Soc. Edinburgh*, sec. A., vol. 61, pp. 186–194, 1942.
2 BARANKIN, E. W.: Concerning Some Inequalities in the Theory of Statistical Estimation, *Skand. Aktuarietidskr.*, vol. 34, pp. 35–40, 1951.
3 BHATTACHARYYA, A.: On Some Analogues of the Amount of Information and Their Use in Statistical Estimation, *Sankhyā, ser. A.*, vol. 8, pp. 1–14, 1946.
4 CHAPMAN, D. G., and H. ROBBINS: Minimum Variance Estimation without Regularity Assumptions, *Ann. Math. Statist.*, vol. 22, pp. 581–586, 1951.

[1] For Probs. 7 to 9 refer to Gart [8].

5 CRAMÉR, H.: "Mathematical Methods of Statistics," Princeton University Press, Princeton, N.J., 1946.

6 FEND, A. V.: On the Attainment of Cramér-Rao and Bhattacharyya Bounds for the Variance of an Estimate, *Ann. Math. Statist.*, vol. 30, pp. 381–388, 1959.

7 FRASER, D. A. S., and I. GUTTMAN: Bhattacharyya Bounds without Regularity Assumption, *Ann. Math. Statist.*, vol. 23, pp. 629–632, 1952.

8 GART, J. J.: An Extension of the Cramér-Rao Inequality, *Ann. Math. Statist.*, vol. 30, pp. 367–380, 1959.

9 KIEFER, J.: On Minimum Variance Estimators, *Ann. Math. Statist.*, vol. 23, pp. 627–629, 1952.

10 Rao, C. R.: Information and Accuracy Attainable in the Estimation of Statistical Parameters, *Bull. Calcutta Math. Soc.*, vol. 37, pp. 81–91, 1945.

11 SETH, G. R.: On the Variance of Estimates, *Ann. Math. Statist.*, vol. 20, pp. 1–27, 1949.

12 WOLFOWITZ, J.: Efficiency of Sequential Estimates, *Ann. Math. Statist.*, vol. 18, pp. 215–230, 1947.

chapter Nine

Maximum-likelihood Estimation

1 □ INTRODUCTION

The method of maximum likelihood is one of the oldest and most important in estimation theory. It was used by Gauss in developing the theory of least squares, which subsequently overshadowed it until it was reintroduced by Fisher [14] in 1912. The method, intimately connected with sufficiency and applicable to almost all problems of estimation, has great practical appeal. Unfortunately, it sometimes produces inferior results for small numbers of observations and even for large sample sizes often results in computational problems. We shall show, however, that the method does possess reasonable large-sample-size properties. We discuss applications of maximum-likelihood methods to problems of reliability inference and compare them with the method of mean-unbiased estimation.

For further work on the subject the reader may refer to Brunk [4] and van Eeden [9, 10], who develop interesting techniques of maximum-likelihood estimation of monotone parameters.

2 □ THE METHOD ILLUSTRATED

Let $x_1, \ldots, x_n$ be observed random variables from a population with a probability distribution indexed by θ. The likelihood function L is defined, in the case where the x_i come from a continuous sample space with probability density $f_\theta(x)$, by

$$L(\theta:x_1, \ldots, x_n) = \prod_{i=1}^{n} f_\theta(x_i) \tag{1a}$$

and for discrete sample spaces by

$$L(\theta:x_1, \ldots, x_n) = \prod_{i=1}^{n} P_\theta(x_i) \tag{1b}$$

Note that the likelihood function is a function of θ for given $x_1, \ldots, x_n$. The method of maximum likelihood consists of finding a value of θ which maximizes the likelihood function. Assuming that the likelihood is a positive differentiable function of θ and that the maximum does not occur on the boundary of the set of all admissible θ, we attempt to find a solution to the likelihood equation

$$\frac{\partial L}{\partial \theta} = 0 \tag{2a}$$

or to the equivalent, but usually simpler, equation

$$\frac{\partial \log L}{\partial \theta} = 0 \tag{2b}$$

(Recall that the logarithm is a monotone function.) We shall ignore any solution which is independent of the observations; i.e., any constant solutions.

EXAMPLE 1

Let $X_1, \ldots, X_n$ be independent random variables, all $N(\theta, \sigma^2)$. Find maximum-likelihood estimates of θ and of σ^2.

$$L(\theta, \sigma^2 : x_1, \ldots, x_n) = (2\pi\sigma^2)^{-n/2} \exp\left[-\sum_{i=1}^{n} \frac{(x_i - \theta)^2}{2\sigma^2}\right]$$

$$\log L = -\sum_{i=1}^{n} \frac{(x_i - \theta)^2}{2\sigma^2} - \frac{n}{2}\log \sigma^2 - \frac{n}{2}\log 2\pi$$

$$\frac{\partial \log L}{\partial \theta} = \frac{1}{\sigma^2}\sum_{i=1}^{n}(x_i - \theta) = 0$$

$$\frac{\partial \log L}{\partial \sigma^2} = \sum_{i=1}^{n} \frac{(x_i - \theta)^2}{2\sigma^4} - \frac{n}{2\sigma^2} = 0$$

Solving these equations, we obtain the maximum-likelihood estimates

$$\theta^* = \sum_{i=1}^{n} \frac{X_i}{n} = \bar{X} \qquad \sigma^{*2} = \frac{1}{n}\sum_{i=1}^{n}(X_i - \bar{X})^2$$

The estimates are exactly those arising from a least-squares fit. This is a particular example of the fact that the maximum-likelihood estimate in the case of normal random variables is the same as the least-squares estimate. Note that while the estimate of θ is unbiased, the estimate of σ^2 is biased, a common feature of maximum-likelihood estimates.

Grouped samples

Kulldorff [26] has investigated the role of maximum likelihood in estimation from grouped and partially grouped samples. Suppose that observations x, independent random variables with common absolutely continuous distribution function $F_\theta(x)$, fall in some interval (z_0, z_n), $-\infty \leq z_0 \leq z_n \leq \infty$, and suppose this interval to be subdivided into n intervals with end points z_0 and z_1, z_1 and z_2, and so on. Partially grouped samples occur when each interval falls into one of two nonvoid sets A and B. For observations falling into any interval in A, the exact value of the observation is recorded, while for intervals in B just the total number of observations per interval is recorded.

We denote the kth observation in the jth interval (where known) by

x_{jk} and the total number of observations in the jth interval by m_j. The probability of an observation falling in the jth interval is

$$p_j = F(z_j) - F(z_{j-1})$$

and the likelihood function L is just

$$L(\theta) = \prod_A \prod_{k=1}^{m_j} f(x_{jk}) \prod_B p_j{}^{m_j}$$

the A denoting multiplication over those intervals in A, the B those in B.

Kulldorff shows in his theorem that under the following four conditions:

(i) $\dfrac{\partial \log p_j}{\partial \theta}, \dfrac{\partial^2 \log p_j}{\partial \theta^2}$ exist for every θ in the parameter space, an open interval, and for every interval in B in which an observation has occurred

(ii) $\dfrac{\partial}{\partial \theta} \log f(x_{jk}), \dfrac{\partial^2}{\partial \theta^2} \log f(x_{jk})$ exist for every interval in A for $k = 1, 2, \ldots, m_j$

(iii) $\displaystyle\sum_A \sum_{k=1}^{m_j} \dfrac{\partial}{\partial \theta} \log f(x_{jk}) + \sum_B m_j \dfrac{\partial \log p_j}{\partial \theta}$

is positive at the left end of the interval and negative at the right end

(iv) $\displaystyle\sum_A \sum_{k=1}^{m_j} \dfrac{\partial^2}{\partial \theta^2} \log f(x_{jk}) + \sum_B \dfrac{\partial^2}{\partial \theta^2} \log p_j$

is negative for any θ satisfying the likelihood equation

$$\sum_A \sum_{k=1}^{m_j} \frac{\partial}{\partial \theta} \log f(x_{jk}) + \sum_B \frac{\partial}{\partial \theta} \log p_j = 0$$

the above likelihood equation has exactly one root, and this root is the maximum-likelihood estimate of θ. (See for proof Kulldorff [26].)

It can be shown that under certain regularity conditions the maximum-likelihood estimate is a consistent and asymptotically efficient estimator (for details see Prob. 12).

Now we give examples to illustrate Kulldorff's theory.

EXAMPLE 2

Let n observations be made on a random variable with density function $f(x) = \alpha e^{-\alpha x}$, $\alpha > 0$. The exact value of each observation less than

154 PARAMETRIC ESTIMATION

M is recorded, and a count is made of all observations greater than or equal to M. Suppose that $x_1, \ldots, x_m$ are recorded and that $n - m$ observation are not less than M. The probability of an observation greater than M is

$$P_M = \int_M^\infty \alpha e^{-\alpha x}\, dx = \int_{\alpha M}^\infty e^{-y}\, dy = e^{-\alpha M}$$

Thus

$$L(\alpha) = \left[\prod_{j=1}^m \alpha e^{-\alpha x_j} \right] (e^{-\alpha M})^{n-m}$$

and also

$$\frac{\partial}{\partial \alpha} \log L = \sum \frac{\partial}{\partial \alpha} \log \alpha e^{-\alpha x_j} + \frac{\partial}{\partial \alpha} [-\alpha M(n-m)]$$

$$= \frac{m}{\alpha} - \sum_{j=1}^m x_j - M(n-m)$$

Setting the likelihood equation equal to zero and solving, we obtain as the maximum-likelihood estimate for α

$$\hat{\alpha} = \frac{m}{\sum_{j=1}^m x_j + M(n-m)}$$

Definition

Let $X_1, \ldots, X_n, \ldots$ be independent random variables with common density function $P_\theta(x)$. An estimate $t_n(X_1, \ldots, X_n)$ is said to be asymptotically efficient in the strict sense if the distribution of $\sqrt{C_n(\theta)} \times (t_n(X_1, \ldots, X_n) - \theta)$ tends to $N(0,1)$ as $n \to \infty$ where

$$C_n(\theta) = E\left[\frac{\partial \log P_\theta(X_1, \ldots, X_n)}{\partial \theta} \right]^2$$

and $P_\theta(x_1, \ldots, x_n)$ is the joint density function of $X_1, \ldots, X_n$.

EXAMPLE 3

The maximum-likelihood estimate of $1/\lambda$ is consistent and asymptotically efficient in the strict sense.

Proof:

In order to prove the result we check the conditions of Prob. 12. Let

$$g_X(x; 1/\lambda) = (2\pi x^3)^{-\frac{1}{2}} \left(\frac{1}{\lambda}\right)^{-\frac{1}{2}} \exp\left[-\frac{1}{2x(1/\lambda)} \right] \quad \text{for } x > 0$$

MAXIMUM-LIKELIHOOD ESTIMATION

Now

$$\ln g_X(x;\lambda) = -\tfrac{1}{2}\ln 2\pi - \tfrac{1}{2}\ln\frac{1}{\lambda} - \frac{3}{2}\ln x - \frac{1}{2x(1/\lambda)}$$

$$\frac{\partial \ln g}{\partial(1/\lambda)} = -\frac{1}{2(1/\lambda)} + \frac{1}{2x(1/\lambda)^2} \quad (i)$$

$$\frac{\partial^2 \ln g}{\partial(1/\lambda)^2} = \frac{1}{2(1/\lambda)^2} - \frac{1}{x(1/\lambda)^3} \quad (ii)$$

$$\frac{\partial^3 \ln g}{\partial(1/\lambda)^3} = \frac{1}{(1/\lambda)^3} + \frac{3}{x(1/\lambda)^4}$$

Therefore

(i) $\dfrac{\partial \ln g}{\partial(1/\lambda)}$ exists for every $\lambda \in (0, \infty)$ and for almost all x.

(ii) $\dfrac{\partial^2 \ln g}{\partial(1/\lambda)^2}$ exists for every $\lambda \in (0, \infty)$ and for almost all x.

(iii) $\dfrac{\partial^3 \ln g}{\partial(1/\lambda)^3}$ exists for every $\lambda \in (0, \infty)$ and for almost all x.

In our case

$$\frac{\partial g}{\partial(1/\lambda)} = (2\pi x^3)^{-\tfrac{1}{2}} \exp\left[-\frac{1}{2x(1/\lambda)}\right]\left(\frac{1}{\lambda}\right)^{-\tfrac{1}{2}}\left[-\frac{1}{2}\left(\frac{1}{\lambda}\right)^{-1} + \frac{1}{2x(1/\lambda)^2}\right]$$

$$\frac{\partial^2 g}{\partial(1/\lambda)^2} = (2\pi x^3)^{-\tfrac{1}{2}} \left(\frac{1}{\lambda}\right)^{-\tfrac{1}{2}} \exp\left[-\frac{1}{2x(1/\lambda)}\right]$$
$$\left[\frac{3}{4}\left(\frac{1}{\lambda}\right)^{-2} + \frac{1}{4x^2(1/\lambda)^4} - \frac{3}{2x(1/\lambda)^3}\right]$$

Therefore

(iv) $\displaystyle\int_0^\infty \frac{\partial g}{\partial(1/\lambda)}\,dx = -\frac{\lambda}{2}\int_0^\infty g\,dx + \frac{\lambda^2}{2}E\left(\frac{1}{X}\right)$

$$= -\frac{\lambda}{2} + \frac{\lambda^2}{2}\frac{1}{\lambda} = 0 \quad \text{for every } \lambda \in (0, \infty)$$

(v) $\displaystyle\int_0^\infty \frac{\partial^2 g}{\partial(1/\lambda)^2}\,dx = \frac{3\lambda^2}{4}\int_0^\infty g\,dx + \frac{\lambda^4}{4}E\left(\frac{1}{X^2}\right) - \frac{3\lambda^3}{2}E\left(\frac{1}{X}\right)$

$$= \frac{3\lambda^2}{4} + \frac{\lambda^4}{4}\frac{3}{\lambda^2} - \frac{3\lambda^3}{2}\frac{1}{\lambda} = 0 \quad \text{for every}$$

$$\lambda \in (0, \infty)$$

Now

$$\int_0^\infty \frac{\partial^2 \ln g}{\partial (1/\lambda)^2} g \, dx = \frac{\lambda^2}{2} \int_0^\infty g \, dx - \lambda^3 E\left(\frac{1}{X}\right)$$

$$= \frac{\lambda^2}{2} - \lambda^3 \frac{1}{\lambda} = -\frac{\lambda^2}{2}$$

Therefore

(vi) $\quad -\infty < \int_0^\infty g \dfrac{\partial^2 \ln g}{\partial (1/\lambda)^2} \, dx < 0$

Let $h(1/\lambda) = (1/\lambda)^2$. Then

$$\frac{\partial^2}{\partial (1/\lambda)^2}\left[\left(\frac{1}{\lambda}\right)^2 \frac{\partial \ln g}{\partial (1/\lambda)}\right] = \frac{\partial^2}{\partial (1/\lambda)^2}\left(-\frac{1}{2}\frac{1}{\lambda} + \frac{1}{2x}\right) = 0$$

Consequently,

(vii) There exists a function $h(1/\lambda)$ which is positive and twice differentiable for every $\lambda \in (0, \infty)$ and a function H (equal to any positive constant) independent of $(1/\lambda)$ such that

$$\left|\frac{\partial^2}{\partial (1/\lambda)^2}\left[h\left(\frac{1}{\lambda}\right)\frac{\partial \ln g}{\partial (1/\lambda)}\right]\right| < H \quad \text{for every } \lambda \in (0, \theta)$$

and such that $\int_0^\infty Hg_X(x, \lambda) \, dx < \infty$.

These seven conditions give the required result.

EXAMPLE 4

The likelihood equation has one and only one root, and this root is the maximum likelihood estimate of $1/\lambda$.

Proof:

In order to prove the result we check the required conditions of the Kulldorff theorem. From Eqs. (*i*) and (*ii*),

(i) $\quad \dfrac{\partial \ln g}{\partial (1/\lambda)}$ and $\dfrac{\partial^2 \ln g}{\partial (1/\lambda)^2}$ exist for every $\lambda \in (0, \infty)$ and for almost all x.

Now

$$\frac{\partial \ln L}{\partial (1/\lambda)} = -\frac{n\lambda}{2} + \frac{\lambda^2}{2}\sum_{i=1}^n \frac{1}{x_i} = 0$$

is the likelihood equation; therefore

(ii) $\left.\dfrac{\partial^2 \ln L}{\partial (1/\lambda)^2}\right|_{1/\lambda = S} = \left[\dfrac{n}{2(1/\lambda)^2} - \dfrac{1}{(1/\lambda)^3} \sum_{i=1}^{n} \dfrac{1}{x_i}\right]\bigg|_{1/\lambda = S}$

$= \dfrac{n^3}{2}\left(\sum_{i=1}^{n} \dfrac{1}{x_i}\right)^{-2} - n^3 \left(\sum_{i=1}^{n} \dfrac{1}{x_i}\right)^{-2}$

$= -\dfrac{n^3}{3}\left(\sum_{i=1}^{n} \dfrac{1}{x_i}\right)^{-2} < 0$

These two conditions together with the solution of the likelihood equation give the result.

Theorem 1

If an estimate attains the Cramér-Rao lower bound, then the likelihood equation has a unique solution.

The proof is assigned as a problem.

Theorem 2

If a sufficient statistic t for a sample from a population with known distribution exists, then any solution of the likelihood equation will be a function of this statistic only.

The proof is assigned as a problem.

On the other hand, Fraser notes that determination of the likelihood equation often produces the relevant sufficient statistics. The first example in this chapter illustrates both these statements. Kendall and Stuart [25] note that given the solution to a maximum-likelihood problem with sufficient statistics, a function of the solution can be found which is an unbiased estimate and has the minimum variance of all such unbiased estimates.

3 ☐ LARGE - SAMPLE - SIZE PROPERTIES

Denoting the solution to the maximum-likelihood equation by $\bar{\theta}(x_1, \ldots, x_n)$, $\bar{\theta}$ is a function of random variables and hence itself is a random variable with parameter θ. The maximum-likelihood method attempts to find the mode of the distribution of $\bar{\theta}$. In estimation theory, the mode is usually inferior to either the mean or the median, and so it is not surprising that the method has poor small-sample-size properties and is often replaced by methods such as the minimax approach. For large sample sizes, however, the mode tends to approach the mean and the median, if they exist, and the method of maximum likelihood has many satisfying aspects for such samples. In fact, the resulting estimate is a consistent, asymptotically normal, and asymptotically efficient estimate of θ for large sample sizes. The proof, the major theorem of this chapter, follows.

Theorem

(i) For almost every x, $\partial/\partial\theta \log f$, $\partial^2/\partial\theta^2 \log f$, and $\partial^3/\partial\theta^3 \log f$ exist for all θ in a nondegenerate interval I.

(ii) For every θ in I, $|\partial f/\partial\theta| < A_1(x)$, $|\partial^2 f/\partial\theta^2| < A_2(x)$, and $|\partial^3/\partial\theta^3 \log f| < A_3(x)$, where $A_1(x)$, $A_2(x)$ are integrable over the real line and

$$\int_{-\infty}^{\infty} A_3(x) f_\theta(x)\, dx < N \qquad N \text{ independent of } \theta$$

(iii) For every θ in I, $\int_{-\infty}^{\infty} \left(\dfrac{\partial}{\partial\theta} \log f\right)^2 f\, dx$ is finite and positive.

Then the likelihood equation, Eq. (2b) of Sec. 2, has a solution converging in probability to θ as n tends to infinity. This solution is an asymptotically efficient estimate of θ; that is, the variance of the estimate tends to the Cramér-Rao lower bound. Furthermore, the solution of Eq. (2b) of Sec. 2 is asymptotically normal.

Proof:

We assume that θ_0, the true value of the parameter, is an interior point of I. We first show that there is a solution of Eq. (2b) of Sec. 2 tending to θ_0 in probability. For every θ in I

$$\frac{\partial}{\partial\theta} \log f = \left(\frac{\partial}{\partial\theta} \log f\right)_{\theta=\theta_0} + (\theta - \theta_0)\left(\frac{\partial^2}{\partial\theta^2} \log f\right)_{\theta=\theta_0} + \tfrac{1}{2}\phi(\theta - \theta_0)^2 A_3(x) \qquad |\phi| < 1 \quad (1)$$

Combining Eqs. (1) with (2b) of Sec. 2, we obtain

$$\frac{1}{n}\frac{\partial}{\partial\theta} \log L = B + C(\theta - \theta_0) + \tfrac{1}{2}\phi' D(\theta - \theta_0)^2 = 0 \qquad (2)$$

where $|\phi'| < 1$ and

$$B = \frac{1}{n} \sum_{i=1}^{n} \left[\frac{\partial}{\partial \theta} \log f_\theta(x_i)\right]_{\theta=\theta_0}$$

$$C = \frac{1}{n} \sum_{i=1}^{n} \left[\frac{\partial^2}{\partial \theta^2} \log f_\theta(x_i)\right]_{\theta=\theta_0} \quad (3)$$

$$\phi'D = \frac{1}{n} \sum_{i=1}^{n} \phi A_3(x)$$

Note that B, C, and D are functions of the random variables. We now must show that for all $\varepsilon > 0$ Eq. (2) has a root θ in the interval $\theta_0 \pm \varepsilon$ with probability tending to 1 as n tends to infinity.

We consider properties of B, C, and D for large n. Since

$$\int_{-\infty}^{\infty} f_\theta(x) \, dx = 1$$

and because of (i) and (ii)

$$\int_{-\infty}^{\infty} \frac{\partial f}{\partial \theta} \, dx = \int_{-\infty}^{\infty} \frac{\partial^2 f}{\partial \theta^2} \, dx = 0$$

for every θ in I, and so

$$E\left(\frac{\partial}{\partial \theta} \log f\right)_{\theta=\theta_0} = \int_{-\infty}^{\infty} \left(\frac{1}{f} \frac{\partial f}{\partial \theta}\right)_{\theta_0} f_{\theta_0}(x) \, dx = 0$$

$$E\left(\frac{\partial^2}{\partial \theta^2} \log f\right)_{\theta=\theta_0} = \int_{-\infty}^{\infty} \left[\frac{1}{f} \frac{\partial^2 f}{\partial \theta^2} - \left(\frac{1}{f} \frac{\partial f}{\partial \theta}\right)^2\right]_{\theta_0} f_{\theta_0}(x) \, dx$$

$$= -E\left(\frac{\partial}{\partial \theta} \log f\right)^2_{\theta_0} = -K^2 \quad (4)$$

where, by (iii), $K > 0$. Since B is the arithmetic mean of n independent random variables of finite variance and zero mean with common distribution, it follows by Khintchine's theorem that B converges to zero in probability. Similarly, it may be shown that C converges in probability to $-K^2$ and D to $E[A_3(x)]$, a nonnegative value less than N.

Let now δ and ε be arbitrarily small given numbers, and let $P(X)$ denote the joint probability density of the random variables, $X_1, \ldots, X_n$. We can choose an $N(\delta,\varepsilon)$ such that for all $n > N(\delta,\varepsilon)$

$$P_1 = P(|B| \geq \varepsilon^2) < \tfrac{1}{3}\delta$$
$$P_2 = P(C \geq -\tfrac{1}{2}K^2) < \tfrac{1}{3}\delta$$
$$P_3 = P(|D| \geq 2N) < \tfrac{1}{3}\delta$$

Let U denote the set of all $\boldsymbol{X} = (X_1, \ldots, X_n)$ such that all three inequalities
$$|B| < \varepsilon^2 \qquad C \leq -\tfrac{1}{2}K^2 \qquad |D| \leq 2N$$
are satisfied. For $\bar{U}$, the complement of the set U,
$$P(\bar{U}) \leq P_1 + P_2 + P_3 < \delta$$
and so
$$P(U) > 1 - \delta$$

Thus $\boldsymbol{X}$ is in $\bar{U}$ with probability at least $1 - \delta$ for $n > N(\delta, \varepsilon)$.

For $\theta = \theta_0 \pm \varepsilon$ the right-hand side of Eq. (2) becomes $B \pm C\varepsilon + \tfrac{1}{2}\phi' D\varepsilon^2$. For every $\boldsymbol{X}$ in U, the sum of the first and last terms is less than $(N + 1)\varepsilon^2$ in absolute value, while
$$C\varepsilon < -\tfrac{1}{2}K^2\varepsilon$$
If $\varepsilon < \tfrac{1}{2}K^2/(N + 1)$, the sign of Eq. (2) at the points $\theta = \theta_0 \pm \varepsilon$ will be determined asymptotically by the second term, so that
$$\frac{\partial}{\partial\theta} \log L > 0 \qquad \text{for } \theta = \theta_0 - \varepsilon$$
and
$$\frac{\partial}{\partial\theta} \log L < 0 \qquad \text{for } \theta = \theta_0 + \varepsilon$$

Furthermore, (i) implies that $\partial/\partial\theta \log L$ is a continuous function of θ for almost all $\boldsymbol{X}$, and so there is a solution to Eq. (2) in the interval $\theta_0 \pm \varepsilon$ with probability greater than $1 - \delta$ for n sufficiently large. Thus the first part of the proof is complete.

Denote by $\theta^* = \theta^*(X_1, \ldots, X_n)$ the solution of Eq. (2). From Eqs. (2) and (3) we obtain

$$K\sqrt{n}(\theta^* - \theta_0) = \frac{\dfrac{1}{K\sqrt{n}} \sum_{i=1}^{n} \left[\dfrac{\partial}{\partial\theta} \log f_\theta(X_i)\right]_{\theta_0}}{-C/K^2 - [\tfrac{1}{2}\phi D(\theta^* - \theta_0)]/K^2} \tag{5}$$

The above work implies that the denominator of the right-hand side of Eq. (5) converges in probability to 1. Since, by Eq. (4), $(\partial/\partial\theta \log f_\theta)_{\theta_0}$ is a function of a random variable with mean zero and standard deviation $-k^2$, then by the Lindeberg-Levy theorem the sum $\sum_{1}^{n} \left[\dfrac{\partial}{\partial\theta} \log f_\theta(X_i)\right]_{\theta_0}$

is asymptotically normal $(0, K\sqrt{n})$ and so the numerator of the right-hand side of Eq. (5) is asymptotically normal (0,1).

Finally, it now follows from the Slutky theorem that $K\sqrt{n}\,(\theta^* - \theta_0)$ is asymptotically normal (0,1), so that θ^* is asymptotically normal $(\theta_0, c/\sqrt{n})$, where

$$c^2 = \frac{1}{K^2} = \frac{1}{E(\partial/\partial\theta \log f_\theta)^2}$$

the Cramér-Rao lower bound, and so the estimate is asymptotically efficient. This completes the proof of the theorem. The proof for the discrete case is similar.

Halperin [20] has investigated the special partial-grouping problem of two intervals, the one containing $-\infty$ being of the first type and the other, containing $+\infty$, being of the second. He has found that with a few more conditions similar to those of the above theorem the maximum-likelihood estimate again has the same large-sample-size properties.

Huzurbazar [22] has shown that under certain regularity conditions, as n becomes sufficiently large, there exists only one consistent maximum-likelihood estimate.

4 □ NUMERICAL SOLUTIONS OF MAXIMUM - LIKELIHOOD EQUATIONS

Although maximum-likelihood equations are often easy to set up in practice, they are often difficult to solve. Iteration techniques can be used, three of which we outline below. The general procedure, starting from an initial estimate $\bar{\theta}_0$ of the true solution, is to calculate successive approximations by

$$\bar{\theta}_{i+1} = \bar{\theta}_i - h(\bar{\theta}_i)\left(\frac{\partial}{\partial\theta}\log L\right)_{\theta=\bar{\theta}_i}$$

Three common choices for $h(\theta)$ are:

(i) $h(\theta) = \left(\dfrac{\partial^2}{\partial\theta^2}\log L\right)^{-1}$

(ii) $h(\theta) = -\left[nE\left(-\dfrac{\partial^2}{\partial\theta^2}\log L\right)\right]^{-1}$, n the sample size.

(iii) $h(\theta) = -k/n$ for some k.

Kale [23], who has investigated the convergence of several iterative processes, points out that only the first method, a Newton-Raphson process, is of second order and so more rapidly convergent for close approximations. Fraser has found that the second method, a first-order process, often provides better convergence for initially poor approximations. Selection of a proper value of k for the third method is often difficult. For a fuller discussion see Deutsch [8], Householder [21], and Kale [23].

5 □ ON RELIABILITY INFERENCE

Reliability theory consists of two main areas of study: modeling and inference. Flehinger [17] is a representative paper on modeling.

We are concerned not with acceptance procedures but with the maximum-likelihood estimate of the reliability and the minimum-variance-unbiased estimate and confidence relations for it.

Let X be a nonnegative random variable, called the *failure time* of the system. Let $F_X(x)$ be the distribution function and $F_C(x) = 1 - F_X(x)$ the complement of the distribution function of X. Let x_m be a positive constant called the *mission time*. The reliability is defined to be the probability that the system survives the mission time, that is,

$$r = P(X > x_m) = F_C(x_m) \tag{1}$$

By a confidence relation depending on a statistic T we mean a real-valued function C_ξ for any observed value ξ of T and any random variable R, $P(R \leq r) = C_\xi(r)$.

We assume that

$$F_X(x) = \begin{cases} 1 - \exp\left[-(x/\theta)^K\right] & \text{if } x \geq 0 \\ 0 & \text{otherwise} \end{cases} \quad \theta, K > 0 \tag{2}$$

This is called the *Weibull distribution*, and it represents failure rates. Note that the conditional density given that $X > x$ is

$$f_X(x|X > x) = \begin{cases} \dfrac{f_X(x)}{1 - F_X(x)} = K\theta^{-K} x^{K-1} & \text{for } x > 0 \\ 0 & \text{otherwise} \end{cases}$$

This conditional-density function is called the *hazard function*. If $K > 1$, the hazard function is increasing and the system is wearing out as time passes. If $K = 1$, the Weibull distribution becomes the exponential distribution, the hazard is constant, and the system is not wearing out. If $K < 1$ the hazard is decreasing; i.e., the longer the system survives, the greater the probability that it will survive a subsequent time interval of specified length. For a highly complex system, $K = 1$ is a good assumption. We shall assume that the value of K is known and give two confidence relations. For detailed discussions of the subject see Tate [35], Pugh [33], Laurent [27], and Basu [2].

An order-statistic confidence relation

Let $X_{(i)}$ be the ith smallest of n independent random variables, each having the Weibull density. Fix i, and let

$$\Lambda = \left(\frac{X_{(i)}}{\theta}\right)^K$$

Lemma 1

The cumulative distribution function of Λ is

$$F_\Lambda(\lambda) = \begin{cases} \int_{e^{-\lambda}}^{1} \frac{n!}{(n-i)!(i-1)!} t^{n-1}(1-t)^{i-1}\, dt & \text{if } \lambda \geq 0 \\ 0 & \text{otherwise} \end{cases}$$

Proof:

The density of $X_{(i)}$ is

$$f_{X_{(i)}}(x) = \begin{cases} \left\{\frac{n!}{(n-i)!(i-1)!}[1 - \exp(-x^K\theta^{-K})]^{i-1}\right\} \times \\ \quad \{K\theta^{-K}x^{K-1}\exp[-(n-i+1)x^K\theta^{-K}]\} & \text{if } x \geq 0 \\ 0 & \text{otherwise} \end{cases}$$

Now $\Lambda = X_i^K \theta^{-K}$,

$$f_\Lambda(\lambda) = \begin{cases} \frac{n!}{(n-i)!(i-1)!}(1-e^{-\lambda})^{i-1}e^{-(n-i+1)\lambda} & \text{if } \lambda \geq 0 \\ 0 & \text{otherwise} \end{cases}$$

164 PARAMETRIC ESTIMATION

Therefore

$$F_\Lambda(\lambda) = \begin{cases} \int_0^\lambda \frac{n!}{(n-i)!(i-1)!}(1-e^{-u})^{i-1} e^{-u(n-i+1)} \, du & \text{if } \lambda \geq 0 \\ 0 & \text{otherwise} \end{cases}$$

Let $t = e^{-u}$, $u = -\ln t$, $du = -dt/t$. Therefore

$$F_\Lambda(\lambda) = \begin{cases} \int_{e^{-\lambda}}^1 \frac{n!}{(n-i)!(i-1)!} t^{n-i}(1-t)^{i-1} \, dt & \text{if } \lambda \geq 0 \\ 0 & \text{otherwise} \end{cases}$$

$$F_\Lambda(\lambda) = P\left[\left(\frac{X_{(i)}}{\theta}\right)^K \leq \lambda\right]$$
$$= P(X_{(i)}^{-K} \lambda \geq \theta^{-K})$$
$$= P\left\{\exp\left[-\lambda\left(\frac{x_m}{X_{(i)}}\right)^K\right] \leq \exp(-x_m^K \theta^{-K})\right\}$$
$$= P(R \leq r) \quad \text{where } R = \exp(-\lambda x_m^K X_{(i)}^{-K})$$

$$C_\lambda(R) = \begin{cases} \int_{e^{-\lambda}}^1 \frac{n!}{(n-i)!(i-1)!} t^{n-i}(1-t)^{i-1} \, dt & \text{if } \lambda \geq 0 \\ 0 & \text{otherwise} \end{cases} \quad (3)$$

A confidence relation based on a sufficient statistic

The Weibull density function is

$$f_X(x) = \begin{cases} K\theta^{-K} x^{K-1} \exp(-x^K \theta^{-K}) & \text{if } x \geq 0 \\ 0 & \text{otherwise} \end{cases} \quad \theta, K > 0$$

K is known, and θ is unknown.
Let $X_1, \ldots, X_n$ be independent random variables, each having the Weibull density. Let

$$\Lambda = \frac{1}{n} \sum_1^n X_i^K \theta^{-K}$$

Lemma 2

The cumulative distribution function of Λ is

$$F_\Lambda(\lambda) = \begin{cases} \int_0^{n\lambda} \frac{1}{(n-1)!} x^{n-1} e^{-x} \, dx & \text{if } \lambda \geq 0 \\ 0 & \text{otherwise} \end{cases} \quad (4)$$

Proof:
Let
$$Y_i = X_i^K \theta^{-K} \qquad y_i = x_i^K \theta^{-K}$$
Then
$$f_{Y_i}(y_i) = \begin{cases} e^{-y_i} & \text{if } y_i \geq 0 \\ 0 & \text{otherwise} \end{cases}$$

The characteristic function of Y_i is

$$\phi Y_i(t) = E(e^{jty_i}) \qquad \text{where } j^2 = -1$$
$$= \int_0^\infty e^{-y_i(1-tj)} \, dy_i$$
$$= (1 - jt)^{-1}$$

Let $Y = \sum_1^n Y_i$. Since the Y_i are mutually independent,

$$\phi_Y(t) = \prod_1^n \phi_{y_i}(t) = (1 - jt)^{-n}$$

Therefore by the uniqueness theorem for characteristic functions, the density of Y is the gamma density with parameters n and 1, that is,

$$f_Y(y) = \begin{cases} \dfrac{1}{(n-1)!} e^{-y} y^{n-1} & \text{if } y \geq 0 \\ 0 & \text{otherwise} \end{cases}$$

Now $\Lambda = (1/n)Y$, $\lambda = (1/n)y$, $dy = n \, d\lambda$; therefore

$$f_\Lambda(\lambda) = \begin{cases} \dfrac{n^n}{(n-1)!} \lambda^{n-1} e^{-n\lambda} & \text{if } \lambda \geq 0 \\ 0 & \text{otherwise} \end{cases} \tag{5}$$

and

$$F_\Lambda(\lambda) = \begin{cases} \dfrac{n^n}{(n-1)!} \int_0^\lambda t^{n-1} e^{-nt} \, dt & \text{if } \lambda \geq 0 \\ 0 & \text{otherwise} \end{cases}$$

Let $x = nt$, $t = x/n$, $dt = dx/n$. Then

$$F_\Lambda(\lambda) = \begin{cases} \int_0^{n\lambda} \dfrac{1}{(n-1)!} x^{n-1} e^{-x} \, dx & \text{if } \lambda \geq 0 \\ 0 & \text{otherwise} \end{cases} \tag{6}$$

Let $\psi = \theta^K \Lambda$. Then

$$F_\Lambda(\lambda) = P(\theta^{-K}\psi \leq \lambda)$$
$$= P\left(\frac{\lambda}{\psi} \geq \theta^{-K}\right)$$
$$= P\left[\exp\left(-\frac{\lambda x_m^K}{\psi}\right) \leq \exp\left(-x_m^K \theta^{-K}\right)\right]$$
$$= P(R \leq r) \quad \text{where } R = \exp\left(-\frac{\lambda x_m^K}{\psi}\right)$$

and

$$C_\lambda(R) = \begin{cases} \int_0^{n\lambda} \frac{1}{(n-1)!} x^{n-1} e^{-x} \, dx & \text{if } \lambda \geq 0 \\ 0 & \text{otherwise} \end{cases} \quad (7)$$

Lemma 3

ψ is a sufficient and complete statistic for r.
The proof is assigned as a problem.

Estimates of r

What is the maximum-likelihood estimate of r? r is a bijective function of θ. Suppose we knew the maximum-likelihood estimate of θ. Could we deduce from this the maximum-likelihood estimate of r?

Lemma 4

Let θ be a parameter of a density function having a maximum-likelihood estimate $\hat{\theta}$, and let $\mu = g(\theta)$ be a bijective function of θ. Then μ has a maximum-likelihood estimate $\hat{\mu}$, and $\hat{\mu} = g(\hat{\theta})$.

Proof:

Let $L(\theta)$ be the likelihood function. Then $L(\theta)$ has a maximum at $\theta = \hat{\theta}$. Since $\mu = g(\theta)$, and g is bijective, therefore $\theta = g^{-1}(\mu)$. Thus the likelihood function can be written $L[g^{-1}(\mu)]$, and it has a maximum at $g^{-1}(\mu) = \hat{\theta}$, that is, at $\mu = g(\hat{\theta})$.

Lemma 5

If K is known, then

$$\hat{\theta} = \left(\frac{1}{n}\sum_1^n x_i^K\right)^{1/K} \qquad (8)$$

Proof:

The joint density of $X_1, \ldots, X_n$ is

$$L = K^n \theta^{-nK} \left(\prod_1^n x_i\right)^{K-1} \exp\left(-\theta^{-K}\sum_1^n x_i^K\right)$$

Therefore

$$\ln L = n \ln K - nK \ln \theta + (K-1)\sum_1^n \ln x_i - \theta^{-K}\sum_1^n x_i^K$$

and

$$\frac{\partial \ln L}{\partial \theta} = -\frac{nK}{\theta} + K\theta^{-K-1}\sum_1^n x_i^K$$

Let $(\partial \ln L)/\partial \theta = 0$. Then

$$\hat{\theta} = \left(\frac{1}{n}\sum_1^n x_i^K\right)^{1/K}$$

$$\left.\frac{\partial^2 \ln L}{\partial \theta^2}\right|_{\theta=\hat{\theta}} = \frac{nK}{\hat{\theta}^2} - K(K+1)\hat{\theta}^{-K-2}\sum_1^n x_i^K$$

$$= \frac{nK}{\hat{\theta}^2}[1 - (K+1)] = -K^2 \frac{n}{\hat{\theta}^2} < 0$$

Thus

$$\hat{\theta} = \left(\frac{1}{n}\sum_1^n x_i^K\right)^{1/K}$$

maximizes the likelihood function. Therefore

$$\hat{r} = \exp\left(-\hat{\theta}^{-K}x_m^K\right) \qquad (9)$$

In every case in which the maximum-likelihood estimate of θ is known, the maximum-likelihood estimate of r is given by (9); e.g., let j be a fixed integer such that $0 < j \leq n$, and let $X_{(1)}, \ldots, X_{(j)}$ be the j smallest out of n independent random variables each having the Weibull density with

168   PARAMETRIC ESTIMATION

the parameter K known. Then

$$\hat{\theta} = \left\{ \frac{1}{j} \left[\left(\sum_{1}^{j} x_{(i)}^{K} \right) + (n - j) x_{(j)}^{K} \right] \right\}^{1/K}$$

and $\hat{r} = \exp(-\hat{\theta}^{-K} x_m^{K})$.

The maximum-likelihood estimate of r is widely used if $K = 1$, but it has the disadvantage of being biased. Bias is a serious disadvantage in the estimation of reliability because of the common practice of estimating the reliability of a system as the product of the reliability estimates of the subsystems, i.e., assuming that the lifetimes of the subsystems are mutually independent. If the estimates of the reliabilities of the subsystems are all biased in the same direction, then the estimate of the reliability of the system will have a large bias. In order to show this effect, we must first find the density function of $\hat{\theta}$. From (5),

$$f_\Lambda(\lambda) = \begin{cases} \dfrac{n^n}{(n-1)!} \lambda^{n-1} e^{-n\lambda} & \text{if } \lambda \geq 0 \\ 0 & \text{otherwise} \end{cases}$$

By (6), $\psi = \theta^K \Lambda$.

$$f_\psi(\psi) = \begin{cases} \dfrac{n^n \theta^{-nK}}{(n-1)!} \psi^{n-1} e^{-n\theta^{-K}\psi} & \text{if } \psi \geq 0 \\ 0 & \text{otherwise} \end{cases} \quad (10)$$

By (8), $\theta = \psi^{1/K}$. Therefore

$$f_{\hat{\theta}}(\hat{\theta}) = \begin{cases} \dfrac{n^n \theta^{-nK} K}{(n-1)!} \hat{\theta}^{nK-1} e^{-n\theta^{-K}\hat{\theta}^K} & \text{if } \hat{\theta} \geq 0 \\ 0 & \text{otherwise} \end{cases}$$

and

$$E(\hat{r}) = \int_0^\infty r(\hat{\theta}) f_{\hat{\theta}}(\hat{\theta}) \, d\hat{\theta}$$

$$= \frac{n^n \theta^{-nK} K}{(n-1)!} \int_0^\infty (\hat{\theta}^{nK-1}) \exp\left(-n\theta^{-K}\hat{\theta}^K - x_m^K \hat{\theta}^{-K}\right) d\hat{\theta}$$

Let $x = \hat{\theta}/\theta$; then $d\hat{\theta} = \theta \, dx$. Therefore

$$E(\hat{r}) = \frac{n^n K}{(n-1)!} \int_0^\infty x^{nK-1} \exp\left(-nx^K - x_m^K x^{-K} \theta^{-K}\right) dx$$

Now assume $K = 1$.

$$E(\hat{r}) = \frac{n^n}{(n-1)!} \int_0^\infty x^{n-1} \exp\left(-nx - \frac{x_m}{x\theta}\right) dx$$

Let $t = nx$. Therefore $x = t/n$, $dx = 1/n\, dt$, and

$$E(\hat{r}) = \frac{1}{(n-1)!} \int_0^\infty t^{n-1} e^{-t} (e^{-x_m/\theta})^{n/t} \, dt$$

$$= \frac{1}{(n-1)!} \int_0^\infty t^{n-1} e^{-t} r^{n/t} \, dt \tag{11}$$

Now suppose a system has reliability 0.951, so that $r = 0.951$, and suppose r is estimated by using the maximum-likelihood estimate $\hat{r}$ with a sample of five items. By (11), the expected value of r is 0.938 (to find this requires considerable calculation because the integral must be evaluated approximately), so that the relative error is 1.4 percent. Now suppose that a complex system consists of 10 of the above systems, whose lifetimes are mutually independent. The reliability of the complex system is $0.951^{10} = 0.605$, the maximum-likelihood estimate of the reliability is $0.938^{10} = 0.527$, and the relative error is 13 percent.

This illustrates the need for an unbiased estimate of r. We shall derive the minimum-variance-unbiased estimate.

Let ψ be a sufficient and complete statistic for r, and let R be an unbiased estimate of r. Then $E(R|\psi)$ is the essentially (using Lebesgue measure) unique minimum-variance-unbiased estimate of r. This is a statement of the Rao-Blackwell theorem.

Let S be defined as the number of X_i's that are strictly greater than x_m, and let $R = (1/n)S$.

Lemma 6

Let the $X_1, \ldots, X_n$ be independent random variables having any density functions with the restriction that $P(X_i > x_m) = r$ for every i. Then R is an unbiased estimate of r.

Proof:

$$E(R) = \frac{1}{n} E(S) = \frac{1}{n} \sum_{i=0}^{n} i \frac{n!}{i!(n-i)!} r^i (1-r)^{n-i}$$

$$= \frac{nr}{n} \sum_{i=1}^{n} \frac{(n-1)!}{(i-1)!(n-i)!} r^{i-1} (1-r)^{n-i}$$

$$= r \sum_{i=0}^{n-1} \frac{(n-1)!}{i!(n-1-i)!} r^i (1-r)^{n-1-i}$$

$$= r(r+1-r)^{n-1} = r$$

Thus $E(R|\psi)$ is the minimum-variance-unbiased estimate of r. Instead of calculating $E(R|\psi)$, we shall write an estimate of r that is a function of ψ and prove that it is unbiased. Since ψ is sufficient and complete for r, this estimate is the minimum-variance-unbiased estimate of r, by the Rao-Blackwell theorem. The estimate of r is

$$r^* = \begin{cases} \left(1 - \dfrac{x_m{}^K}{n\psi}\right)^{n-1} & \text{if } \psi > \dfrac{x_m{}^K}{n} \\ 0 & \text{otherwise} \end{cases} \quad (12)$$

Lemma 7

$\lim_{n \to \infty} r^* = r$, the maximum-likelihood estimate of r given by (9).

Proof:

This is a special case of the fact that if x, a, and b are any fixed real numbers, then

$$\lim_{n \to \infty} \left(1 + \frac{x}{n-a}\right)^{n-b} = e^x$$

Before proving the unbiasedness of r^*, we need a result about the incomplete gamma function.

Lemma 8

If $C \geq 0$, then

$$\int_C^\infty x^{n-1} e^{-x} \, dx = e^{-C}(n!) \sum_0^{n-1} \frac{C^i}{i!} \quad (13)$$

Proof:

Induction on n. Assume that Eq. (13) is true for n

$$\int_C^\infty x^n e^{-x} \, dx = C^n e^{-C} + n \int_C^\infty x^{n-1} e^{-x} \, dx$$

Integrating by parts,

$$\int_C^\infty x^n e^{-x} \, dx = e^{-C} \left[C^n + n\overline{|(n)|} \sum_0^{n-1} \frac{C^i}{i!} \right] \quad \text{by (12)}$$

$$= e^{-C}[\Gamma(n+1)] \sum_0^n \frac{C^i}{i!}$$

Theorem

r^* is unbiased.

Proof:

$$Er^* = \int_{x_m^K/n}^{\infty} \left(1 - \frac{x_m^K}{n}\psi\right)^{n-1} f_\psi(\psi)\,d\psi$$

$$= \int_{x_m^K/n}^{\infty} \sum_{0}^{n-1} \binom{n-1}{i}\left(\frac{-x_m^K}{n}\psi\right)^i$$

$$\frac{n^n\theta^{-nK}}{(n-1)!}\psi^{n-1}\exp(-n\psi\theta^{-K})\,d\psi \quad \text{by (10)}$$

$$= \sum_{0}^{n-1} \frac{n^n\theta^{-nK}}{i!(n-1-i)!}(-1)^i x_m^{Ki} n^{-i} \int_{x_m^K/n}^{\infty} \psi^{n-1-i}\exp(-n\psi\theta^{-K})\,d\psi$$

If in the last integral we let $x = n\theta^{-K}\psi$, we obtain

$$\theta^{nK-iK}n^{i-n}\int_{x_m^K\theta^{-K}}^{\infty} x^{n-1-i} e^{-x}\,dx$$

$$= \theta^{nK-iK}n^{i-n}(n-1-i)!\exp(-x_m^K\theta^{-K})\sum_{j=0}^{n-1-i}\frac{x_m^{jK}\theta^{-jK}}{j!} \quad \text{by Lemma 8}$$

Therefore

$$Er^* = \exp(-x_m^K\theta^{-K})\sum_{i=0}^{n-1}\frac{(-1)^i}{i!}\theta^{-Ki}x_m^{Ki}\sum_{j=0}^{n-1-i}\frac{x_m^{jK}\theta^{-jK}}{j!}$$

$$= r\sum_{i=0}^{n-1}\sum_{j=0}^{n-1-i}(-1)^i\frac{(x_m^K\theta^{-K})^{i+j}}{i!j!}$$

Let $S = i + j$ so that $j = S - i$.

$$Er^* = r\sum_{S=0}^{n-1}(x_m^K\theta^{-K})^S\sum_{i=0}^{S}\frac{(-1)^i}{i!(S-i)!}$$

$$= r\sum_{S=0}^{n-i}(x_m^K\theta^{-K})^S\frac{(1-1)^S}{S!}$$

$$= r$$

For another application of maximum-likelihood estimation method to a problem of reliability see Prob. 17.

PROBLEMS

1. For $x_1, \ldots, x_n$ independent random variables with the following probability distributions, set up the likelihood equations, solve, and determine sufficient statistics where possible. Are the maximum-likelihood estimates minimum-variance unbiased?

(a) $P(x = 0) = q$ $\quad q$ unknown; $P(x = 1) = p = 1 - q$.

(b) A Poisson distribution of unknown mean λ.

(c) A gaussian distribution of unknown mean μ and of known variance σ_0^2; of known mean μ_0 and unknown variance σ^2.

(d) $x_i = e^{y_i}$, y_i from a gaussian distribution of unknown mean and variance. Find the estimates for the mean and variance of the x_i.

(e) $x_1, \ldots, x_m$ from a population with distribution $N(\mu, \sigma^2)$; $x_{m+1}, \ldots, x_n$ from a population with distribution $N(\mu, \gamma\sigma^2)$. Find the maximum-likelihood estimate of μ.

(f) A gamma distribution of unknown mean and variance.

(g) A beta distribution.

(h) A Cauchy distribution.

(i) A distribution uniform over $(0, \theta)$, $\theta > 0$, and zero elsewhere.

(j) A distribution with the density function

$$f_\theta(x) = \frac{\theta^{j+1} x^j e^{-\theta x}}{\Gamma(j+1)} \quad \begin{array}{l} \text{for fixed } j \\ \theta, x > 0 \end{array}$$

2. Let $x_1, \ldots, x_n$ be independent random variables with density function

$$P(x = 0) = 1 - p$$
$$P(x = 1) = p$$

where $0 \leq p \leq \frac{1}{2}$. Find the maximum-likelihood estimate of p.

3. For the above population, but with $0 \leq p \leq 1$, samples are drawn until the sum of the samples is n. At this point m samples have been drawn. Show that the distribution of m is given by

$$\binom{m-1}{n-1} p^n (1-p)^{m-n} \quad m = n, n+1, \ldots$$

and the maximum-likelihood estimate of p is n/m.

4. Given $X_1, \ldots, X_n$ independent random variables with density function

$$f(x) = \exp\left(-\sum_{i=0}^{m} a_i x^i\right)$$

show that the method of moments agrees with the method of maximum likelihood in estimating the a_i, $i = 1, \ldots, m$.

5. Prove Kulldorff's theorem stated in Sec. 2.

6. (Kulldorff) For a partially grouped sample from a population with density function

$$f(x) = \mu e^{-\mu x} \qquad \mu > 0$$

find the maximum-likelihood estimate of μ.

7. For a partially grouped sample from a population of independent normal random variables with known variance, find the maximum-likelihood estimate of the mean. Then assume that the mean is known and the variance unknown; find its maximum-likelihood estimate.

8. A random sample of N objects is drawn from a population with normal distribution; n_1 fall below a certain known standard, and n_2 fall above a second, higher known standard. Calculate the maximum-likelihood estimates of the mean and the variance. Calculate the distribution of the maximum-likelihood estimate of the mean.

9. Prove Theorem 1 of Sec. 2.

10. Prove Theorem 2 of Sec. 2.

11. For the density of part (j) of Prob. 1 show that the asymptotic distribution of the maximum-likelihood estimate for large n is $N[\theta, \theta^2/n(j+1)]$.

12. Assume the setup of Sec. 2 and given the following assumptions, prove that the maximum-likelihood estimate of θ is a consistent estimate of θ and is asymptotically efficient in the strict sense:

(i) $\dfrac{\partial \log P_i}{\partial \theta}$ exists for every $\theta \in \Omega_0$, for $i = 1, \ldots, k$.

(ii) $\dfrac{\partial^2 \log P_i}{\partial \theta^2}$ exists for every $\theta \in \Omega_0$, for $i = 1, \ldots, k$.

(iii) $\dfrac{\partial^3 \log P_i}{\partial \theta^3}$ exists for every $\theta \in \Omega_0$, for $i = 1, \ldots, k$.

(iv) $\sum_{i=1}^{k} \dfrac{\partial P_i}{\partial \theta} = 0$ for $\theta = \theta_0$.

(v) $\sum_{i=1}^{k} \dfrac{\partial^2 P_i}{\partial \theta^2} = 0$ for $\theta = \theta_0$.

(vi) $-\infty < \sum_{i=1}^{k} P_i \dfrac{\partial^2 \log P_i}{\partial \theta^2} < 0$ for $\theta = \theta_0$.

(vii) There exists a function $g(\theta)$ which is positive and twice differentiable for every $\theta \in \Omega_0$ and a set of numbers H_i (independent of θ but possibly dependent on θ_0) such that

(a) $\left| \dfrac{\partial^2}{\partial \theta^2} \left[g(\theta) \dfrac{\partial \log P_i}{\partial \theta} \right] \right| < H_i \quad$ for every $\theta \in \Omega_0$
$$i = 1, \ldots, k$$

(b) $\sum_{i=1}^{k} p_i(\theta) H_i < \infty$.

13. (Epstein and Sobel) Let X be a random variable with density function

$$f_0(x) = \begin{cases} \dfrac{1}{\theta} \exp\left(-\dfrac{x - A}{\theta} \right) & \text{for } x \geq A > 0 \\ & \theta > 0 \\ 0 & \text{for } x < A \end{cases}$$

Suppose a random sample of N items is drawn from the population of this density and tested. Let the items be divided into K sets, each set containing n_j items. Suppose the items in each set are observed only until the first r_j failures occur. r_j and n_j are fixed and preassigned and strictly positive, $r_j \leq n_j$. Consider the following cases:

(i) The items in each set have a common known A in the density.
(ii) All N items have a common unknown A.
(iii) The items in each set have a common unknown A.

(a) Find the unique minimum-variance-unbiased estimate $\hat{\theta}$ of θ for each case.

(b) Find the maximum-likelihood estimate θ^* of θ and show it to be proportional to that of the unique minimum-variance-unbiased estimate of θ.

(c) Let R be the total number of failures and find distribution of $2R\hat{\theta}/\theta$ for each case.

(d) What does A represent?

(e) What does θ represent?

14. (Barton) Let $X_1, X_2, \ldots, X_n$ be independent random variables normally distributed with mean θ and variance 1. Then show that

$$\hat{R}(t,\hat{\theta}) = (2\pi)^{-\frac{1}{2}} \int_{[n/(n-1)]^{-\frac{1}{2}}(t-\bar{X})}^{\infty} e^{-x^2/2}\, dx$$

is the minimum-variance-unbiased estimate of

$$R(t,\theta) = \int_{t}^{\infty} \frac{1}{\sqrt{2\pi}} e^{-(x-\theta)^2/2}\, dx \qquad \text{where } -\infty < \theta < \infty$$

and

$$\bar{X} = \frac{1}{n}\sum_{1}^{n} X_i$$

15. (Barton) Let $X_1, X_2, \ldots, X_n$ be independent random variables normally distributed with mean θ and variance σ^2 (both unknown). Then prove that

$$\hat{R}(t,\hat{\theta},\hat{\sigma}^2) =$$

$$\begin{cases} 1 & \text{if } t < \bar{X} - S(n-1)^{\frac{1}{2}} \\ (n-1)^{-\frac{1}{2}} \dfrac{1}{B[\frac{1}{2},(n-2)/2]} \\ \qquad \int_{(t-\bar{X})/S}^{(n-1)^{\frac{1}{2}}} \left(1 - \dfrac{x^2}{n-1}\right)^{(n-4)/2} dx & \text{if } \bar{X} - S(n-1)^{\frac{1}{2}} \le t \le \bar{X} + S(n-1)^{\frac{1}{2}} \\ 0 & \text{if } t > \bar{X} + S(n-1)^{\frac{1}{2}} \end{cases}$$

where

$$\bar{X} = \frac{1}{n}\sum_{1}^{n} X_i, \qquad \text{and} \qquad S^2 = \frac{1}{n}\sum_{i=1}^{n}(X_i - \bar{X})^2$$

is the minimum-variance-unbiased estimate of

$$R(t,\theta,\sigma^2) = \int_{t}^{\infty} \frac{1}{\sqrt{2\pi\sigma^2}} e^{-(x-\theta)^2/2\sigma^2}\, dx \qquad \begin{array}{l} -\infty < \theta < \infty \\ 0 < \sigma^2 < \infty \end{array}$$

16. (Tate) Let $X_1, X_2, \ldots, X_n$ be independent random variables with density function

$$f(x,\theta) = \begin{cases} \dfrac{x^{p-1}}{\theta^p \overline{|(p)|}} e^{-x/\theta} & \text{for } x \geq 0,\ \theta > 0 \\ & p \text{ a known integer} \\ 0 & \text{otherwise} \end{cases}$$

Then prove

$$\hat{R}(t,\hat{\theta}) = \begin{cases} 1 - \dfrac{1}{\beta[p,(n-1)p]} \displaystyle\int_0^{t/\bar{X}} x^{p-1}(1-x)^{(n-1)p-1}\,dx & \text{if } 0 \leq t \leq \bar{X} \\ 0 & \text{if } t > \bar{X} \end{cases}$$

where $\bar{X} = \dfrac{1}{n}\displaystyle\sum_1^n X_i$ is the minimum-variance-unbiased estimate of

$$R(t,\theta) = \int_t^\infty \dfrac{x^{p-1}}{\theta^p \overline{|p|}} e^{-x/\theta}\,dx \qquad \begin{array}{l}\theta > 0 \\ p \text{ a known integer}\end{array}$$

17. (Mendenhall and Hader) We suppose that there are two subpopulations, subpopulation 1 being present in the proportion p, and subpopulation 2 being present in the proportion q, where $p + q = 1$. The subpopulation indicates the cause of failure.

Let X represent the failure time, as usual. We assume that the cumulative distribution function of the ith subpopulation ($i = 1$ and 2) is

$$F_i(x) = \begin{cases} 1 - e^{-x/\theta_i} & \text{for } x \geq 0 \\ 0 & \text{otherwise} \end{cases} \qquad (1)$$

where $\theta_i > 0$. Then the density function is

$$f_i(x) = \begin{cases} \dfrac{1}{\theta_i} e^{-x/\theta_i} & \text{for } x \geq 0 \\ 0 & \text{otherwise} \end{cases} \qquad (2)$$

The cumulative distribution function of the whole population is

$$F(x) = pF_1(x) + qF_2(x)$$

The density function is

$$f(x) = pf_1(x) + qf_2(x)$$

Let $G_i = 1 - F_i$ and $G = 1 - F$. If every item in the population were put on test, the proportion of items belonging to each subpopulation would, in general, change with time, since items in one subpopulation will fail sooner than items in the other. Let $p(x)$ and $q(x)$ be the proportion of items in subpopulations 1 and 2 at time x. Clearly $p(0) = p$ and $p(x) = pG_1(x)/G(x)$.

A random sample of n items is drawn and placed on test. At a preassigned time t the test is stopped, and r items have failed, r_i from the ith subpopulation. The $n - r$ items which have survived the test cannot be assigned to one subpopulation or the other. Let X_{ij} be the time of the jth failure from the ith subpopulation.

Assume that θ_1/θ_2 is unknown.
Let $u = X/t$ and $\beta_i = \theta_i/t$. The cumulative distribution function of u is

$$F_i(u) = \begin{cases} 1 - e^{-u/\beta_i} & \text{for } u \geq 0 \\ 0 & \text{otherwise} \end{cases} \quad (3)$$

Given a sample of size n, the probability of r_i items failing from cause i is

$$P(r_1, r_2) = \begin{cases} \dfrac{n!}{r_1! r_2! (n - r)!} [pF_1(1)]^{r_1} [qF_2(1)]^{r_2} [G(1)]^{n-r} \\ \qquad\qquad\qquad\qquad\qquad\qquad \text{if } r_1 + r_2 \leq n \quad (4) \\ 0 \qquad\qquad\qquad\qquad\qquad\qquad \text{otherwise} \end{cases}$$

where $r = r_1 + r_2$, the F_i are given by (3), and

$$G(u) = 1 - pF_1(u) - qF_2(u)$$

The conditional density of $u_{i1}, \ldots, u_{ir_i}$, given that $u_{i1} \leq \cdots \leq u_{ir_i} \leq 1$ and that $R_i = r_i$, is

$P(r_1, r_2)$

$$= \begin{cases} \dfrac{n!}{(n-r)!} G^{n-r}(1) p^{r_1} q^{r_2} \prod_{j=1}^{r_1} f_1(u_{ij}) \prod_{j=1}^{r_2} f_2(u_{2j}) \\ \qquad\qquad\qquad\qquad\qquad \text{if } u_{i1} \leq \cdots \leq u_{ir_i} \leq 1 \quad (5) \\ \qquad\qquad\qquad\qquad\qquad r_1 + r_2 \leq n \\ 0 \qquad\qquad\qquad\qquad\qquad \text{otherwise} \end{cases}$$

Call this function L.

(a) Find a maximum-likelihood estimate of β_1, β_2, p, and K, where $K = pG_1(1)/G(1)$.

(b) Let β_1/β_2 be known to be ≤ 1 or ≥ 1. Suppose it is known that $\beta_1 \leq \beta_2$. If $\hat{\beta}_1 > \hat{\beta}_2$, it is said that a crossover has occurred;

assume that $\beta_1 = \beta_2 = \beta$. Then show that $G(1) = e^{-1/\beta}$ and

$$L = \frac{n!}{(n-r)!} e^{-(n-r)/\beta} p^{r_1} q^{r_2} \prod_{i=1}^{r_1} f(\mu_{1j}) \prod_{i=1}^{r_2} f(\mu_{2j})$$

Find a maximum-likelihood estimate of β and p.

(c) Define the information matrix for the parameters β_1, β_2, and p, consider their maximum-likelihood estimates, and discuss their optimum properties.

18. (Brunk and van Eeden) Let $f_n(\theta)$ be a function unimodal in θ with a unique maximum at θ_n^*, $n = 1, 2, \ldots$, if the product $\Pi f_i(\theta)$ of any finite number of these functions is unimodal with a unique maximum, then prove that $(\hat{\theta}_1, \ldots, \hat{\theta}_n)$ maximizes $\prod_{K=1}^{n} f_k(\theta_K)$ subject to $0 \leq \theta_1 \leq \cdots \leq \theta_n$ if

$$\hat{\theta}_K = \max_{1 \leq \alpha \leq k} \min_{k \leq \beta \leq n} M(\alpha, \beta)$$

where $M(\alpha, \beta)$ is the maximizing value of $\prod_{k=\alpha}^{\beta} f_k(\theta)$.

REFERENCES

1 BARTON, D. E.: Unbiased Estimation of a Set of Probabilities, *Biometrika*, vol. 48, pp. 227–229, 1961.

2 BASU, A. P.: Estimates of Reliability for Some Distributions Useful in Life Testing, *Technometrics*, vol. 6, pp. 215–219, 1964.

3 BOSEWELL, M. T.: Estimating and Testing Trend in a Stochastic Process of Poisson Type, *Ann. Math. Statist.*, vol. 37, pp. 1564–1573, 1966.

4 BRUNK, H. D.: Maximum Likelihood Estimates of Monotone Parameters, *Ann. Math. Statist.*, vol. 26, pp. 607–616, 1955.

5 BRUNK, H. D., W. R. FRANK, D. L. HANSON, and R. V. HOGG: Maximum Likelihood Estimation of the Distributions of Two Stochastically Ordered Random Variables, *J. Am. Statist.*, vol. 61, pp. 1067–1080, 1966.

6 CRAMÉR, H.: "Mathematical Methods of Statistics," Princeton University Press, Princeton, N.J., 1946.

7 Davis, D. J.: An Analysis of Some Failure Data, *J. Am. Statist. Assoc.*, vol. 47, pp. 113–150, 1952.

8 DEUTSCH, R.: "Estimation Theory," Prentice-Hall, Inc., Englewood Cliffs, N.J., 1965.

9 EEDEN, C. van: Maximum Likelihood Estimation of Partially or Completely Ordered Parameters, I and II, *Koninkl. Ned. Akad., ser. A,* Van Wet Amsterdam, vol. 60, pp. 128–136 and 201–211, 1957.

10 EEDEN, C. VAN: Maximum Likelihood of Ordered Probabilities, *Koninkl. Ned. Akad. Wetenschap. Proc., ser. A,* vol. 89, pp. 444–455, 1956.

11 EEDEN, C. VAN: A Least Squares Inequality for Maximum Likelihood Estimates of Ordered Parameters, *Koninkl. Ned. Akad., ser. A,* Van Wet Amsterdam, vol. 60, pp. 513–521, 1957.

12 EPSTEIN, B., and M. SOBEL: Some Theorems Relevant to Life Testing for an Exponential Distribution, *Ann. Math. Statist.,* vol. 25, pp. 373–381, 1954.

13 EPSTEIN, B., and M. SOBEL: Sequential Life Tests in the Exponential Case, *Ann. Math. Statist.,* vol. 26, pp. 82–93, 1955.

14 FISHER, R. A.: On an Absolute Criterion for Fitting Frequency Curves, *Messsenger of Math.,* vol. 41, p. 155, 1912.

15 FISHER, R. A.: On the Mathematical Foundation of Theoretical Statistics, *Phil. Trans. Royal Soc. London, ser. A,* vol. 222, pp. 309–368, 1922.

16 FISHER, R. A.: Theory of Statistical Estimation, *Proc. Cambridge Phil. Soc.* vol. 22, pp. 700–725, 1925.

17 FLEHINGER, B. J.: A General Model for the Reliability Analysis of System under Preventive Maintenance Policies, *Ann. Math. Statist.,* vol. 33, pp. 137–156, 1962.

18 FRASER, D. A. S.: "Nonparametric Methods in Statistics," John Wiley & Sons, Inc., New York, 1957.

19 FRASER, D. A. S.: "The Structure of Inference," John Wiley & Sons, Inc., New York, 1968.

20 HALPERIN, M.: Maximum Likelihood Estimation in Truncated Samples, *Ann. Math. Statist.,* vol. 23, pp. 226–238, 1952.

21 HOUSEHOLDER, A. S.: "Principles of Numerical Analysis," McGraw-Hill Book Company, New York, 1953.

22 HUZURBAZAR, V. S.: The Likelihood Equation, Consistency and the Maxima of the Likelihood Function, *Ann. Eugen,* vol. 14, pp. 185–200 (1948).

23 KALE, B. R.: On the Solution of the Likelihood Equation by Iteration Processes, *Biometrika,* vol. 48, pp. 452–456, 1961.

24 KENDALL, M. G.: "Exercises in Theoretical Statistics," Charles Griffin & Company, Ltd., London, 1954.

25 KENDALL, M. G., and A. STUART: "The Advanced Theory of Statistics," vol. 2, Charles Griffin & Company, Ltd., London, 1961.

26 KULLDORFF, G.: "Contributions to the Theory of Grouped and Partially Grouped Samples," Almqvist and Wiksell, Stockholm, 1961.

27 LAURENT, A. G.: "Conditional Distribution of Order Statistics and Distribution of the Reduced *i*th Order Statistics of the Exponential Model," *Ann. Math. Statist.,* vol. 34, pp. 652–657, 1963.

28 LECAM, L.: On the Asymptotic Theory of Estimation and Testing Hypotheses, *Proc. 3rd Berkeley Symp. Math. Statist. Probability,* pp. 129–156, 1956.

29 LEHMANN, E. L.: "Testing Statistical Hypotheses," John Wiley & Sons, Inc., New York, 1959.

30 LOEVE, M.: "Probability Theory," D. Van Nostrand, Company, Inc., Princeton, N.J., 1960.

31 MARSHALL, A. W., and F. PROSCHAN: Maximum Likelihood Estimation for Distributions with Monotone Failure Rate, *Ann. Math. Statist.*, vol. 36, pp. 69–77, 1965.

32 MENDENHALL, W., and R. J. HADER: Estimation of Parameters of Mixed Exponentially Distributed Failure Time Distributions from Censored Life Test Data, *Biometrika*, vol. 45, pp. 504–520, 1958.

33 PUGH, E. L.: The Best Estimate of Reliability in the Exponential Case, *Operation Res.*, vol. 11, pp. 57–61, 1963.

34 RAO, C. R.: "Linear Statistical Inference and Its Applications," John Wiley & Sons, Inc., New York, 1965.

35 TATE, R. F.: Unbiased Estimation: Functions of Location and Scale Parameters, *Ann. Math. Statist.*, vol. 30, pp. 341–366, 1959.

36 THOMPSON, W. A., JR.: The Problem of Negative Estimates of Variance Components," *Ann. Math. Statist.*, vol. 33, pp. 273–289, 1962.

37 WILKS, S. S.: "Mathematical Statistics," John Wiley & Sons, Inc., 1962.

chapter *Ten*

Admissible and Minimax Estimation

1 □ INTRODUCTION

In this chapter, admissible, Bayes, and minimax estimation procedures are discussed, and their relations are established. Constant-risk, Cramér-Rao inequality, and Bayes methods of construction of a minimax estimate of a parameter are demonstrated with examples. As in other chapters, we restrict ourselves to univariate density functions.

The relation of minimax estimation theory to the theory of games is discussed for the special case of a binomial parameter in Sec. 4. A comparison of minimax estimation with unbiased estimation reveals that minimaxity is suited to small samples. We discuss the properties of Bayes estimates and the relation of the invariance principle to minimaxity and admissibility.

The risk function $R_\delta(\theta) = E_\theta\{w[\theta, \delta(X)]\}$ was used in the previous sections to judge the desirability of an estimator $\delta(x)$ of the parameter θ, the criterion being the smaller the risk, the better the estimator. However, in the absence of some uniformly minimum risk function, the work in the previous chapters has been restricted to the "respectable" subset of estimators which are unbiased. In this chapter a different approach is taken, and an optimum property of the risk function itself will be found, in this case the minimum supremum.

Definition   Minimax Estimate

If a random variable X has a density function $p_\theta(x)$ and $\delta(X)$ is some estimate of θ, then the risk function is $R_\delta(\theta) = E_\theta\{w[\theta,\delta(X)]\}$. A minimax estimator $\delta(X)$ is any estimator which minimizes the $\sup_\theta R_\delta(\theta)$. Although the use of a minimax estimate is not always wise, there are many cases where it is useful. For example, if a certain error can be tolerated over a bounded range but an error outside this range would be catastrophic, a minimax estimator would be an excellent choice.

2 □ ADMISSIBLE ESTIMATE

The concept of admissibility helps in quickly ruling some estimators out of consideration. Thus it should lead to some sort of ordering of estimators. We give now below definitions which help to understand this concept.

Definition 1

An estimate δ is said to be *as good as* δ_0 (where both δ and δ_0 are estimates of θ) if $R_\delta(\theta) \leq R_{\delta_0}(\theta)$ for all $\theta \in \Omega$.

Definition 2

An estimate δ is said to be *better than* δ_0 if $R_\delta(\theta) \leq R_{\delta_0}(\theta)$ for all $\theta \in \Omega$ and $R_\delta(\theta) < R_{\delta_0}(\theta)$ for at least one $\theta \in \Omega$. When $R_\delta(\theta) = R_{\delta_0}(\theta)$ for all $\theta \in \Omega$, then δ and δ_0 are said to be *equivalent*.

Definition 3

An estimate δ of θ is said to be *admissible* if there exists no estimate of θ better than δ. Thus if δ and δ_0 are two estimates of θ and $R_\delta(\theta) \leq R_{\delta_0}(\theta)$

184   PARAMETRIC ESTIMATION

with strict inequality for at least one value of $\theta \in \Omega$, then δ_0 is said to be inadmissible. Obviously an inadmissible estimator cannot be an estimator which minimizes the supremum of $R_\delta(\theta)$. Note, however, that the inadmissibility of δ_0 does not imply the admissibility of $\delta(X)$.

Definition 4   Essentially Complete Family of Estimates

A family ζ of estimates of θ is said to be essentially complete if, given any estimate δ_0 not in ζ, there exists an estimate δ in ζ that is as good as δ_0.

Definition 5   Complete Family of Estimates

A family η of estimates of θ is said to be complete if, given any estimate δ_0 not in η, there exists an estimate δ in η that is better than δ_0.

Depending on what the risk function is minimized with respect to, minimax estimation may take on several forms. Often, one of the easiest ways to minimize it is to use a priori probability distribution. This form of estimation is known as *Bayes estimation*.

Definition 6   Bayes Estimate

Let us consider a probability measure m on a σ-field of subsets of Ω. Then a Bayes estimate δ_m is defined as one which minimizes the average risk relative to a priori probability measure m. That is,

$$\int_\Omega R_{\delta_m}(\theta)\, dm \leq \int_\Omega R_\delta(\theta)\, dm \qquad \text{for all } \delta$$

Bayes estimation is related to the minimax procedure by the following theorems, which are due to Hodges and Lehmann [5].

3 □ THEOREMS AND THEIR APPLICATIONS

Theorem 1

An estimator δ_m which is a Bayes estimator relative to the measure m and for which R_{δ_m} is a constant over Ω is minimax.

Proof:
If δ is any estimator,

$$\sup_{\theta \in \Omega} R_\delta(\theta) \geq \int_\Omega R_\delta(\theta)\, dm$$
$$\geq \int_\Omega R_{\delta_m}(\theta)\, dm$$
$$= K = \text{const}$$
$$= \sup_{\theta \in \Omega} R_{\delta_m}(\theta)$$

then

$$\sup_{\theta \in \Omega} R_\delta(\theta) \geq \sup_{\theta \in \Omega} R_{\delta_m}(\theta)$$

Method of finding the Bayes estimate

Let X be a random variable with density function $p_\theta(x)$, where $\theta \in \Omega$. Let $\delta(X)$ be an estimate of $f(\theta)$, a function of θ. Let the squared error $[\delta(x) - f(\theta)]^2$ be the loss function. The Bayes estimate δ_π corresponding to a priori distribution π is the estimate δ which minimizes

$$\int_\Omega R_\delta(\theta)\, d\pi(\theta) = \int_\Omega \int_{R_1} [\delta(x) - f(\theta)]^2 p_\theta(x)\, dx\, d\pi(\theta)$$
$$= \int_{R_1} \int_\Omega [\delta(x) - f(\theta)]^2 p_\theta(x)\, d\pi(\theta)\, dx$$

The interchange of the order of integration is valid by the Fubini theorem (the conditions of which we assume). Thus we have to look at the inner integral

$$\int_\Omega [\delta(x) - f(\theta)]^2 p_\theta(x)\, d\pi(\theta)$$
$$= [\delta(x)]^2 \int_\Omega p_\theta(x)\, d\pi(\theta) - 2\delta(x) \int_\Omega f(\theta) p_\theta(x)\, d\pi(\theta) + \int_\Omega [f(\theta)]^2 p_\theta(x)\, d\pi(\theta)$$

This is minimized for

$$\delta(x) = \frac{\int_\Omega f(\theta) p_\theta(x)\, d\pi(\theta)}{\int_\Omega p_\theta(x)\, d\pi(\theta)}$$

Thus this is the Bayes estimate $\delta_\pi(x)$.

Theorem 2

If δ is a minimax estimator for some set $\Omega_0 \subset \Omega$ and

$$\sup_{\theta \in \Omega_0} R_\delta(\theta) = \sup_{\theta \in \Omega} R_\delta(\theta)$$

then δ is a minimax estimator for Ω.

The proof is left to the reader.

Theorem 3

If δ is a minimax estimate for Ω and $\sup_{\theta \in \Omega} R_{\delta*}(\theta) = \sup_{\theta \in \Omega_0} R_{\delta*}(\theta)$ for all δ^*, where $\Omega_0 \subset \Omega$, then δ is a minimax estimator for Ω_0.

The proof is left to the reader.

Careful study of several detailed examples will make the meaning of these theorems clearer and will bring out some important properties of minimax estimators.

EXAMPLE 1

If X is a random variable whose density function is given by the binomial distribution (n,p), find a minimax estimate of p when the loss function is the squared error.

The distribution is $p(x) = \binom{n}{x} p^x q^{n-x}$, and the loss is $[\delta(x) - p]^2$, where $\delta(X)$ is the estimator. Assume that $\delta(X)$ is a linear estimate of the form $aX + b$, where a and b are constants. Then

$$\begin{aligned}
R_\delta(p) &= E_p[(aX + b - p)^2] \\
&= E_p\{[a(X - np) + b + (an - 1)p]^2\} \\
&= a^2npq + [b + (an - 1)p]^2 \\
&= [(an - 1)^2 - a^2n]p^2 + [a^2n + 2b(an - 1)]p + b^2
\end{aligned}$$

This is a quadratic equation in p. To find the values of a and b for which the risk is a constant, set the coefficients of p and p^2 equal to zero and solve the resulting simultaneous equations.

$$(an - 1)^2 - a^2n = 0$$
$$a^2n + 2b(an - 1) = 0$$

$$a = \frac{1}{\sqrt{n}\,(1 + \sqrt{n})} \quad \text{or} \quad \frac{1}{\sqrt{n}\,(1 - \sqrt{n})}$$

$$b = \frac{1}{2(1 + \sqrt{n})} \quad \text{or} \quad \frac{1}{2(2n - 3\sqrt{n} - 1)}$$

The second set of roots must be discarded since the resulting estimator could be outside the interval [0,1], a known impossibility. Now the estimator has the form

$$\delta(x) = \frac{x}{\sqrt{n}\,(1+\sqrt{n})} + \frac{1}{2(1+\sqrt{n})}$$

If an a priori distribution $\pi(p)$ can be found for which the Bayes estimator has constant risk, a minimax estimator will have been found. Consider

$$\delta_\pi(x) = \frac{\int_0^1 p \binom{n}{x} p^x q^{n-x} \pi(p)\, dp}{\int_0^1 \binom{n}{x} p^x q^{n-x} \pi(p)\, dp}$$

The solution technique is to assume a density $\pi(p)$ and test it to see whether it satisfies the above criterion. Assume a density

$$\pi(p) = c p^{d-1}(1-p)^{\beta-1}$$

the beta distribution with $d > 0$, $\beta > 0$, It is then found that

$$\delta_\pi(x) = \frac{\int_0^1 p^{d+x}(1-p)^{n+\beta-x-1}\, dp}{\int_0^1 p^{d+x-1}(1-p)^{n+\beta-x-1}\, dp}$$

Using the commonly known result that

$$\int_0^1 x^{n-1}(1-x)^{m-1} = \frac{\Gamma(m)\Gamma(n)}{\Gamma(m+n)}$$

the above estimate becomes

$$\delta_\pi(x) = \frac{\Gamma(d+x+1)\Gamma(n+\beta-x)\Gamma(d+\beta+n)}{\Gamma(x+d)\Gamma(n+\beta-x)\Gamma(d+\beta+n+1)}$$

$$= \frac{x}{n+d+\beta} + \frac{d}{n+d+\beta}$$

For this to be identically equal to $ax + b$ for all x,

$$a = \frac{1}{n+d+\beta} = \frac{1}{\sqrt{n}\,(1+\sqrt{n})}$$

$$b = \frac{d}{n+d+\beta} = \frac{1}{2(1+\sqrt{n})}$$

which is true if $d = \beta = \sqrt{n}/2$. Therefore the estimate

$$\delta = \frac{\sqrt{n}}{1+\sqrt{n}}\frac{x}{n} + \frac{1}{2(1+\sqrt{n})}$$

is a minimax estimator and has constant risk function

$$R_{(p)} = \frac{1}{4(1+\sqrt{n})^2}$$

EXAMPLE 2

Let $x_1, x_2, \ldots, x_n$ be independent random variables with common distribution function F. Let Ω be the set of all distribution functions F such that $P_F(0 \leq x_1 \leq 1) = 1$, and let Ω_0 be the set of all distribution functions F such that $0 < P_F(x_1 = 1) < 1$ and

$$P_F(x_1 = 0) = 1 - P_F(x_1 = 1).$$

If the loss function is squared error, we know from the previous problem that

$$\delta(x) = \frac{\sqrt{n}}{1+\sqrt{n}}\bar{x} + \frac{1}{2(1+\sqrt{n})}$$

is a minimax estimate for Ω_0. Show that $R_\delta(F) = 1/4(1+\sqrt{n})^2$ for $F \in \Omega$,

$$\begin{aligned}R_\delta(F) &= E_F\left\{\frac{\sqrt{n}}{1+\sqrt{n}}[\bar{x} - E_F(\bar{x})] + \frac{1}{2(1+\sqrt{n})} - \frac{E_F(x_1)}{1+\sqrt{n}}\right\}^2 \\ &= \frac{1}{(1+\sqrt{n})^2}\operatorname{var}_F x_1 + \frac{1}{(1+\sqrt{n})^2}\left[\frac{1}{2} - E_F(x_1)\right]^2 \\ &= \frac{1}{(1+\sqrt{n})^2}\{E_F(x_1^2) - E_F^2(x_1) + [\tfrac{1}{2} - E_F(x_1)]^2\} \\ &= \frac{1}{4}\frac{1}{(1+\sqrt{n})^2}\end{aligned}$$

Use Theorem 2 to assert that δ is minimax for Ω. Hence it may be concluded that if the loss function is square error, a, b are real numbers and δ is a minimax estimator for θ; then $a\delta + b$ is a minimax estimator for $a\theta + b$.

EXAMPLE 3

Given two independent random variables x and y with binomial distributions (n,p_1) and (n,p_2), find a minimax estimate for $p_1 - p_2$ assuming a squared-error loss function.

One's first inclination is to try to solve the problem using a linear estimator in a manner similar to the other binomial minimax problem. However, it would be found that no linear minimax estimate of $p_1 - p_2$ exists. Now the technique is to attempt to find a linear estimate over some subset. Pick the subset $\Omega = \{p_1 = p, \ p_2 = 1 - p\}$ and define a new variable $z = x + (n - y)$ whose binomial distribution is $(2n,p)$; the estimate $p_1 - p_2$ becomes $2p - 1$. The previous example says that a Bayes estimate of p which has constant risk over Ω_0 is

$$\frac{1}{1 + \sqrt{2n}} \left(\frac{z}{\sqrt{2n}} + \frac{1}{2} \right)$$

Therefore a Bayes estimate of $2p - 1$ which has constant risk over Ω_0 is

$$\frac{2}{1 + \sqrt{2n}} \left(\frac{z}{\sqrt{2n}} + \frac{1}{2} \right) - 1 = \frac{\sqrt{2n}}{1 + \sqrt{2n}} \frac{x - y}{n}$$

Work often involves sequences of distributions in a space. The following lemma is particularly useful in finding minimax estimators in these cases.

Lemma

For a given sequence of distributions $p_1, p_2, \ldots$ in the space Ω let $\delta_n = \delta_{p_n}$ be the Bayes solution and $r_n = \int R_{\delta_n}(\theta) \, dp_n(\theta)$ be the Bayes risk. If $\lim_{n \to \infty} r_n \to r$ and δ is any estimate having risk $R_\delta(\theta) = r$, δ is a minimax estimator.

Proof:

$$\sup R_{\delta_0}(\theta) \geq \int_\Omega R_{\delta_0}(\theta) \, dp_n(\theta) \geq \int_\Omega R_{\delta_n}(\theta) \, dp_n(\theta) = r_n$$

where δ_0 is any estimate. Therefore

$$\sup_\theta R_{\delta_0}(\theta) \geq r = \sup R_\delta(\theta)$$

Hence δ is a minimax estimator.

EXAMPLE 4

If $X_1, X_2, \ldots, X_n$ are n independent gaussian random variables with common density function $N(\theta,1)$ and the loss function is the squared error, find a minimax estimator of the mean θ.

No least favorable distribution of θ exists since such a distribution would have to be uniform over $(-\infty, \infty)$. Therefore consider a sequence of a priori distributions $p_\sigma = N(0,\sigma^2)$, with $\sigma = 1, 2, \ldots$. If θ has an a priori distribution p_σ, then

$$p(\theta|x_1,x_2,\ldots,x_n) = \frac{p(\theta,x_1,x_2,\ldots,x_n)}{p(x_1,x_2,\ldots,x_n)}$$

For $n = 1$ and $x = x_1$

$$p(\theta|x) = \frac{N(x,\theta,1)N(\theta,0,\sigma^2)}{\int_{-\infty}^{\infty} N(x,\theta,1)N(\theta,0,\sigma^2)\,d\theta}$$

$$= \frac{1}{\sqrt{2\pi\sigma^2/(1+\sigma^2)}} \exp\left[-\frac{1}{2}\frac{1+\sigma^2}{\sigma^2}\left(\theta - \frac{\sigma^2}{1+\sigma^2}x\right)^2\right]$$

Furthermore

$$\delta_\sigma(x) = E(\theta|x) = \frac{x\sigma^2}{1+\sigma^2}$$

and

$$\operatorname{var}(\theta|x) = \frac{\sigma^2}{1+\sigma^2}$$

Then

$$\sup_\sigma \operatorname{var}(\theta|x) = \sup_\sigma \frac{\sigma^2}{1+\sigma^2} = 1$$

and

$$\lim_{\sigma \to \infty} \delta_\sigma(x) = \lim_{\sigma \to \infty} \frac{\sigma^2 x}{1+\sigma^2} = x$$

Therefore X is minimax.

To show that X is admissible, assume the opposite and develop a contradiction. If X is not admissible, there exists a δ such that

$$R_\delta(\theta) \leq R_x(\theta) = 1$$

With this assumption there exists a number $\varepsilon > 0$ and other numbers

c and d ($c < d$) such that $R_\delta(\theta) < 1 - \varepsilon$ for $\theta \in [c,d]$. Then

$$1 \geq \underset{\delta}{\text{var}}\,(\theta|x) = \int_\Omega R_\delta(\theta)\,dp_\sigma(\theta) \geq \int R_{\delta_\sigma}(\theta)\,dp_\sigma(\theta)$$

$$= \text{var}\,(\theta|x) = \frac{\sigma^2}{1 + \sigma^2}$$

and

$$1 - \underset{\delta}{\text{var}}\,(\theta|x) = \int_{-\infty}^{\infty} [1 - R_\delta(\theta)] N(\theta, 0, \sigma^2)\,d\theta$$

$$\geq \int_c^d [1 - R_\delta(\theta)] N(\theta, 0, \sigma^2)\,d\theta$$

$$\geq \frac{\varepsilon}{\sigma\sqrt{2\pi}} \int_c^d \exp\left(-\frac{\theta}{2\sigma^2}\right) d\theta$$

$$\lim_{\sigma \to \infty} \sigma[1 - \underset{\delta}{\text{var}}\,(\theta|x)] \geq \frac{\varepsilon}{\sqrt{2\pi}}\,(d - c)$$

and

$$\lim_{\sigma \to \infty} \sigma[1 - \text{var}\,(\theta|x)] = \lim_{\sigma \to \infty} \sigma\left(1 - \frac{\sigma^2}{1 + \sigma^2}\right) = 0$$

Hence $\text{var}_\delta\,(\theta|x) < \text{var}\,(\theta|x)$, which is a contradiction, and therefore it is concluded that X is admissible.

NOTE

We include the following section here because for the given stopping rule the situation becomes nonsequential.

*4 □ BINOMIAL MINIMAX ESTIMATION FOR A GIVEN STOPPING RULE

Let $x_1, x_2, \ldots, x_n, \ldots$ be a sequence of independent random variables with common density function

$$P(x = 1) = p$$
$$P(x = 0) = 1 - p \qquad 0 \leq p \leq 1$$

We shall consider a nonrandomized sequential procedure for estimating p or a function of p.

In this section we consider a problem of minimax estimation. For a given stopping rule, the existence of minimax and admissible estimate is proved, and some examples are constructed for small sample sizes to demonstrate the existence theorem. Since minimaxity is a property of small samples, they themselves are of some interest. Furthermore, example 1 shows that the a priori distribution leading to the minimax estimate is not unique.

The following nomenclature is used. A region R is a set of points containing $(0,0)$. The point (n_2,u_2) is immediately beyond (n_1,u_1) if either $n_2 = n_1 + 1$, $u_2 = u_1$ or $n_2 = n_1 + 1$, $u_2 = u_1 + 1$. A path in R from α_0 to the point α_n is a finite sequence of points $\alpha_0, \alpha_1, \ldots, \alpha_n$ such that $\alpha_i (i > 0)$ is immediately beyond α_{i-1}, and $\alpha_i \in R$ with the possible exception of α_n. A boundary point, i.e., an element of the boundary B of R, is a point not in R which is the last point α_n of a path from the origin. The accessible points are the points in R which can be reached by paths from the origin, while inaccessible points are the points which cannot be reached by any path from the origin. n, the first coordinate, is called the *index* of a point, and the index of a region is the least upper bound of the indices of its accessible points. A finite region is a region for which the indices of the accessible points are less than some number n.

A path may be thought of as arising by a random process such that a path reaching $\alpha_i = (n,u)$, $\alpha_i \in R$, will be extended to $\alpha_i = (n+1, u+1)$ with probability p or to $\alpha_{i+1} = (n+1, u)$ with probability $q = 1 - p$. When a path is extended to a boundary point of R, the process ceases. The probability that a boundary point or an accessible point will be included in a path from the origin is $P(\alpha) = K(\alpha) p^u q^{n-u}$, where $K(\alpha)$ is the number of paths from the origin to the point. We call $P(\alpha)$ the *probability* of the point. Any region for which $\sum_{\alpha \in B} P(\alpha) = 1$ is called *closed*. The stopping rule is denoted by δ, which must say, for every possible $n, x_1, \ldots, x_n$, whether to observe x_{n+1} or to stop.

Theorem

Let δ be a closed and bounded stopping rule with r stopping points (n_i,u_i), $i = 1, \ldots, r$, and let $(Z_\delta - p)^2$ be a loss function, where $0 \leq Z_\delta \leq 1$ is an estimate of p. Then there exists the unique minimax, hence admissible, estimate of p and a least favorable distribution corresponding to which the estimate is Bayes.

Proof:
The expected loss will be

$$R(p,Z) = \sum_{\substack{(n_i,u_i)\\i=1}}^{r} [z(n_i,u_i) - p]^2 P_r(n_i,u_i)$$

where $0 \leq z(n_i,u_i) \leq 1$, $i = 1, 2, \ldots, r$. We can transfer this problem into the terms of game theory for the problem of a two persons-zero-sum game. Let the two persons be nature (player I) and the statistician (player II). Nature's strategy is $p \in [0,1] = P$. The statistician's strategy is

$$Z = [z(n_1,u_1), z(n_2,u_2), \ldots, z(n_i,u_i), \ldots, z(n_r,u_r)]$$

a point in the r-dimensional space whose cartesian product is

$$Z = [0,1] \times [0,1] \cdots [0,1].$$

(i) A payoff is an expected loss, that is, $R(p,Z)$, $p \in [0,1]$ and $Z \in Z$.
(ii) $R(p,Z)$ for fixed p is convex in Z . . . .
(iii) $R(p,Z)$ is continuous in p for fixed Z and continuous in Z for fixed p, because it is a polynomial in p and Z. . . .
(iv) Thus $G = (P,Z,R)$ is a game where $P = [0,1]$ is a closed bounded convex subset of S_r . . . , where S_r is the cartesian r-dimensional space of all vectors $S_r = (s_1, \ldots, s_r)$, each s_i, $i = 1, 2, \ldots, r$ being a real number.

We now appeal to the following results of game theory given in [1, p. 51]. "If $G = (P,Z,R)$ is a game such that Z is a closed bounded convex subset of S_r and $R(p,Z)$ is for each p a continuous convex function of Z, then G has a value, player II has a good pure strategy and, for every $\varepsilon > 0$, player I has an ε-good strategy which is a mixture of at most $m + 1$ pure strategies." The conditions of this theorem are satisfied, as shown in (ii), (iii), and (iv). Thus there exists Z_0 such that

$$\inf_{Z \in Z} \sup_{p \in P} R(p,Z) = \sup_{p \in P} R(p,Z_0)$$

Hence there exists a minimax estimate of p. Then we appeal to the following result of [1, p. 53]: "If in addition P is a closed bounded subset of S_1 and R is for each Z continuous in p, then I has a good strategy which is a mixture of at most $m + 1$ pure strategies." Since the conditions of this result are satisfied, there exists an a priori distribution which is least favorable. The existence of a good strategy (for player I) which is a mixture of at most $m + 1$ pure strategies is equivalent to the existence of a discrete a priori distribution of p, where at most $m + 1$ points of $p \in [0,1]$ can have positive probability. It follows (see [1, prob. 2.5.1]) that the minimax estimate is unique and it is player II's good pure strategy. Thus there exists the unique minimax (hence admissible) estimate of p and a least favorable distribution.

Corollary

Under the same hypothesis of the above theorem except that we have loss function

$$(Z - p)^2 + CN$$

where $C > 0$ is a real number, there exists the unique minimax estimate of p and a least favourable distribution.

Proof:

The result follows immediately from the results of game theory mentioned for the above section.

EXAMPLE 1

Stopping point	Probability of stopping point	Constant-risk estimate of p taking squared error $(Z - p)^2$ as loss function
(2,2)	p^2	$\dfrac{2\sqrt{2} + 1}{2(\sqrt{2} + 1)}$
(2,0)	q^2	$\dfrac{1}{2(\sqrt{2} + 1)}$
(3,2)	$2p^2q$	$\frac{1}{2}$
(3,1)	$2pq^2$	$\frac{1}{2}$

And the constant risk is $1/4(1 + \sqrt{2})^2$.

We try the following a priori distribution $\lambda(p)$, corresponding to which the estimate turns out to be Bayes.

P	$\lambda(p)$
0	$\dfrac{1}{2(1+\sqrt{2})}$
$\dfrac{1}{2}$	$\dfrac{2\sqrt{2}}{2(1+\sqrt{2})}$
1	$\dfrac{1}{2(1+\sqrt{2})}$

Then the constant-risk estimate is a minimax estimate, and it is unique, which follows from the theorem. Hence it is admissible.

EXAMPLE 2

Stopping point	Probability of stopping point	Constant-risk estimate of p taking $(Z-p)^2$ as loss function
(3,3)	p^3	0.82365
(3,0)	q^3	0.17635
(4,3)	$3p^3q$	0.61443
(4,2)	$6p^2q^2$	0.50000
(4,1)	$3pq^3$	0.38557

We try the following a priori distribution $\lambda(p)$:

P	$\lambda(p)$
0	0.1295
0.32	0.3705
0.68	0.3705
1.00	0.1295

Since corresponding to this a priori distribution the constant-risk estimate turns out to be Bayes, it is minimax and admissible with constant risk = 0.0311.

EXAMPLE 3

Stopping point	Probability of stopping point	Constant-risk estimate of p taking $(Z-p)^2$ as loss function
(4,4)	p^4	0.84076
(4,0)	q^4	0.15924
(5,4)	$4p^4q$	0.676
(5,3)	$10p^3q^2$	0.5704
(5,2)	$10p^2q^3$	0.4296
(5,1)	$4pq^4$	0.324

We try the following a priori distribution $\lambda(p)$

P	$\lambda(p)$
0.02	0.117367
0.314	0.382633
0.686	0.382633
0.98	0.117367

Since corresponding to this a priori distribution the constant-risk estimate turns out to be Bayes, it is minimax and admissible with constant expected loss 0.02535738.

EXAMPLE 4

Stopping point	Probability of stopping point	Constant-risk estimate of p taking $(Z-p)^2 + CN$ as loss function
(2,2)	p^2	$\dfrac{3 - \sqrt{2+C}}{2}$
(2,0)	q^2	$\dfrac{\sqrt{2+C} - 1}{2}$
(3,2)	$2p^2q$	$\tfrac{1}{2}$
(3,1)	$2pq^2$	$\tfrac{1}{2}$

We try the following a priori distribution $\lambda(p)$:

P	$\lambda(p)$
0	$\dfrac{2 - \sqrt{2+C}}{2\sqrt{2+C}}$
$\dfrac{1}{2}$	$\dfrac{2(\sqrt{2+C} - 1)}{\sqrt{2+C}}$
1	$\dfrac{2 - \sqrt{2+C}}{2\sqrt{2+C}}$

Since corresponding to this a priori distribution the constant-risk estimate turns out to be Bayes, it is minimax and admissible with constant risk

$$\frac{3 + C - 2\sqrt{2+C}}{4}$$

5 □ MINIMAX ESTIMATES BY THE CRAMÉR - RAO INEQUALITY METHOD

The Cramér-Rao inequality can be used to find admissible minimax estimates (see Hodges and Lehmann [4] for a detailed discussion).

EXAMPLE 1

If $x_1, x_2, \ldots, x_n$ are independent gaussian random variables with common density function $N(\mu,1)$ and the loss function is the squared error, let us use the Cramér-Rao inequality to find a minimax estimate $\delta(x)$ for the mean μ. Since the loss function is convex, only nonrandomized estimates need be considered. If the expected value of the estimate is

$$E[\delta(x_1, x_2, \ldots, x_n)] = \mu + b(\mu)$$

the Cramér-Rao inequality is

$$\sigma_\delta^2(\mu) \geq \frac{[1 + b'(\mu)]^2}{n}$$

and the risk function is

$$R_\delta(\mu) = E\{[\delta(x) - \mu]^2\} = E(\{\delta(x) - E[\delta(x)] + b(\mu)\}^2)$$
$$= \sigma_\delta^2(\mu) + b^2(\mu)$$

Using these results,

$$R_\delta(\mu) \geq b^2(\mu) + \frac{[1 + b'(\mu)]^2}{n} \quad \text{for all } \mu \tag{1}$$

It is known that the estimate $\bar{x}$ has the risk function $1/n$. We now prove that this is minimax. The fact that the risk function of δ as a function of μ is less than or equal to $1/n$ implies that

$$b^2(\mu) + \frac{[1 + b'(\mu)]^2}{n} \leq \frac{1}{n}$$

Since both terms on the left-hand side of this equation are nonnegative, $b(\mu)$ is bounded and $b^2(\mu) \leq 1/n$. Because of the nonnegativeness of the terms,

$$\frac{[1 + b'(\mu)]^2}{n} \leq \frac{1}{n}$$
$$1 + 2b'(\mu) + [b'(\mu)]^2 \leq 1$$
$$2b'(\mu) + [b'(\mu)]^2 \leq 0$$

which implies that $b'(\mu) \leq 0$; hence, $b(\mu)$ is nonincreasing. Since $b(\mu)$ is bounded, there exists a sequence of $\{\mu_i\}_i$ approaching infinity for which $b'(\mu_i) \to 0$, and by the same argument there exists a sequence $\{\mu_i\} \to -\infty$ for which $b'(\mu) \to 0$. The convergence of the two sequences to zero implies that $b^2(\mu_i) \to 0$ and $b(\mu_i) \to 0$. Hence since $b(\mu)$ is bounded and nonincreasing, $b(\mu) \equiv 0$ for all μ. Therefore the inequality of Eq. (1) becomes $R \geq 1/n$.

Since $R_\delta(\mu) \leq 1/n$ implies $R_\delta(\mu) \geq 1/n$, it may be concluded that $R_0(\mu) = 1/n$ for all μ. Therefore $\bar{x}$ is an admissible minimax estimate. It is also a unique estimate from the completeness property of $\bar{x}$.

EXAMPLE 2

Let X/θ be a variate having the density function chi-square with n degrees of freedom; then $X/(n + 2)$ is admissible and minimax for loss $(\delta - \theta)^2/\theta^2$.

Proof:

$$E\left(\frac{X}{n+2} - \theta\right)^2 = E\left(\frac{X - n\theta}{n+2} - \frac{2\theta}{n+2}\right)^2$$

$$= \frac{2n\theta^2}{(n+2)^2} + \frac{4\theta^2}{(n+2)^2}$$

$$= \frac{2\theta^2}{n+2}$$

$$R_{x/n+2}(\theta) = \frac{2}{n+2}$$

$$P_\theta(x) = \frac{x^{n/2-1}}{2^{n/2}\left|\left(\frac{n}{2}\right)\right|\theta^{n/2}} e^{-x/2\theta}$$

$$\frac{d}{d\theta}\log P_\theta(x) = \frac{d}{d\theta}\left(-\frac{x}{2\theta} + \frac{n}{2}\log\theta\right)$$

$$= \frac{x}{2\theta^2} - \frac{n}{2\theta}$$

$$E\left[\frac{d}{d\theta}\log P_\theta(X)\right]^2 = E\left(\frac{X}{2\theta^2} - \frac{n}{2\theta}\right)^2$$

$$= E\left(\frac{X - n\theta}{2\theta^2}\right)^2$$

$$= \frac{2n\theta^2}{4\theta^4}$$

$$= \frac{n}{2\theta^2}$$

Let δ be an estimate such that

$$E_\theta \delta = r(\theta) = \theta + b(\theta)$$
$$E(\delta - \theta)^2 = E_\theta[\delta - r(\theta) + b(\theta)]^2$$
$$= \text{var}_\theta\, \delta + b^2(\theta)$$
$$\geq b^2(\theta) + \frac{[1 + b'(\theta)]^2}{E[d/d\theta \log P_\theta(X)]^2}$$

Suppose δ is better than $X/(n+2)$.

$$\frac{2}{n+2} \geq \frac{E_\theta(\delta - \theta)^2}{\theta^2} \geq \frac{b^2(\theta)}{\theta^2} + \frac{[1 + b'(\theta)]^2}{n/2}$$

$$\frac{b^2(\theta)}{\theta^2} \leq \frac{2}{n+2} < 1$$

$$\left| \frac{b(\theta)}{\theta} \right| < 1$$

$$|b(\theta)| < \theta \qquad \lim_{\substack{\theta \to 0 \\ \theta > 0}} b(\theta) = 0$$

$$\frac{2}{n+2} \geq \frac{2}{n}[1 + b'(\theta)]^2$$

$$1 > \frac{n}{n+2} \geq [1 + b'(\theta)]^2$$

$$1 \geq |1 + b'(\theta)| \geq 1 + b'(\theta)$$

$$0 > b'(\theta)$$

Therefore $b(\theta)$ is decreasing in θ.

To show that

$$b'(\theta) \leq \frac{b(\theta)}{\theta} \qquad \theta > 0$$

assume that

$$b'(\theta) > \frac{b(\theta)}{\theta}$$

$$1 + b'(\theta) > 1 + \frac{b(\theta)}{\theta} > 0$$

$$\frac{2}{n+2} \geq \frac{b^2(\theta)}{\theta^2} + \frac{2}{n}[1 + b'(\theta)]^2$$

$$> \frac{b^2(\theta)}{\theta^2} + \frac{2}{n}\left[1 + \frac{b(\theta)}{\theta}\right]^2$$

$$\geq \min_{-\infty < z < \infty} \left[z^2 + \frac{2}{n}(1+z)^2\right]$$

$$= \frac{4}{(n+2)^2} + \frac{2}{n}\left(1 - \frac{2}{n+2}\right)^2$$

$$= \frac{2}{(n+2)^2}(2+n) = \frac{2}{n+2}$$

which is not possible.

Therefore $b'(\theta) \leq b(\theta)/\theta$.

$$\frac{d[b(\theta)/\theta]}{d\theta} = \frac{\theta b'(\theta) - b(\theta)}{\theta^2} = \frac{b'(\theta)}{\theta} - \frac{b(\theta)}{\theta^2} = \frac{1}{\theta}\left[b'(\theta) - \frac{b(\theta)}{\theta}\right] \leq 0$$
$$\text{for } \theta > 0$$

Hence $b(\theta)/\theta$ is decreasing in θ.

To show that $b'(\theta)$ is not bounded away from $b(\theta)/\theta$ for large θ, suppose otherwise there is $\varepsilon > 0$

$$b'(\theta) < \frac{b(\theta)}{\theta} - \varepsilon \qquad \theta \geq \theta_0$$

Let

$$C(\theta) = -\varepsilon \theta \log \theta + K\theta \qquad K = \frac{b(\theta_0) + \varepsilon \theta_0 \log \theta_0}{\theta_0}$$

$$C(\theta_0) = b(\theta_0)$$
$$C'(\theta) = -\varepsilon \log \theta + K - \varepsilon$$
$$= \frac{C(\theta)}{\theta} - \varepsilon$$

$$C'(\theta) - b'(\theta) > \frac{C(\theta) - b(\theta)}{\theta} \qquad \theta \geq \theta_0$$

$$\frac{\theta[C'(\theta) - b'(\theta)] - [C(\theta) - b(\theta)]}{\theta^2} > 0 \qquad \theta \geq \theta_0$$

$$\left[-\frac{C(\theta) - b(\theta)}{\theta}\right]' > 0$$

$$\frac{C(\theta) - b(\theta)}{\theta} = 0 \qquad \text{if } \theta = \theta_0$$
$$> 0 \qquad \text{if } \theta > \theta_0$$

Therefore

$$C(\theta) \geq b(\theta) > -\theta \qquad \theta \geq \theta_0$$
$$\frac{C(\theta)}{\theta} > -1$$

Thus $-\varepsilon \log \theta + K > -1$ for all large θ, and so $-\infty > -1$, a contradiction.

To show $b'(\theta)$ is not bounded away from $b(\theta)/\theta$ for small θ

$$b'(\theta) < \frac{b(\theta)}{\theta} - \varepsilon \qquad \theta \leq \theta_0$$

Define $C(\theta)$ as before

$$\frac{C(\theta) - b(\theta)}{\theta} < 0 \qquad \theta < \theta_0$$

$$\frac{C(\theta)}{\theta} < \frac{b(\theta)}{\theta} < 0 \qquad \text{for } \theta < \theta_0$$

Thus $-\varepsilon \log \theta + K < 0$ for all small θ, and so $+\infty < 0$, a contradiction.

Thus again there is contradiction. Hence $b'(\theta)$ is not bounded away from $b(\theta)/\theta$.

$$\left[\frac{b(\theta)}{\theta}\right]^2 + \frac{2}{n}\left\{1 + \frac{b(\theta)}{\theta} - \left[\frac{b(\theta)}{\theta} - b'(\theta)\right]\right\}^2 = \left[\frac{b(\theta)}{\theta}\right]^2$$
$$+ \frac{2}{n}\left[1 + \frac{b(\theta)}{\theta}\right]^2 - \frac{4}{n}\left[1 + \frac{b(\theta)}{\theta}\right]\left[\frac{b(\theta)}{\theta} - b'(\theta)\right] + \frac{2}{n}\left[\frac{b(\theta)}{\theta} - b'(\theta)\right]^2$$

$\theta_1, \theta_2, \ldots$ is a sequence such that $b(\theta_i)/\theta_i - b'(\theta_i) \to 0$ as $i \to \infty$.

$$\frac{2}{n+2} \geq \lim \left\{\left[\frac{b(\theta_i)}{\theta_i}\right]^2 + \frac{2}{n}\left[1 + \frac{b(\theta_i)}{\theta_i}\right]^2\right\} \geq \frac{2}{n+2}$$

since

$$z^2 + \frac{2}{n}[1 + z]^2 > \frac{2}{n+2} \qquad \text{if } z \neq \frac{-2}{n+2}$$

$$\lim_{i \to \infty} \frac{b(\theta_i)}{\theta_i} = -\frac{2}{n+2} \qquad \text{as } \theta \to 0 \qquad b(\theta) = \frac{-2\theta}{n+2}$$

a contradiction, and the proof is complete.

6 □ BAYES ESTIMATE FOR QUADRATIC LOSS FUNCTION

Let the sample space be (Z,Ω,p) and loss function $L(\omega,a) = \lambda(\omega)[\theta(\omega) - \alpha]^2$, $\lambda(\omega) > 0$, where $\omega \in \Omega$ and $a \in A$. For any function r on Ω and any $\xi \in E$ we shall denote by $E_\xi(r|Z)$ the conditional expected value of r given Z, where ξ is the a priori distribution on Ω. Thus by fixing a and setting $r = L$, we get

$$T_Z(a) = E_\xi[\lambda(\theta - a)^2|Z]$$

dropping the argument ω.

Theorem 1 (*Girschick and Savage* [3])

For each Z, $T_Z(a) < \infty$ either for no a, for exactly one a, or for every $a \in A$. The second case implies that $E_\xi(\lambda|Z) = \infty$ and the third that $E_\xi(\lambda|Z) < \infty$.

Proof:

We first show that if $T_z(a)$ is finite for two distinct points, say a_1, a_2 with $a_1 < a_2$ in A, then it is finite for all $a \in A$.

The function $\lambda(\theta - a)^2$ is convex in a so that for all a in the closed interval $[a_1, a_2]$ and all ω

$$\lambda(\omega)[\theta(\omega) - a]^2 \leq \frac{a_2 - a}{a_2 - a_1} \lambda(\omega)[\theta(\omega) - a_1]^2 + \frac{a - a_1}{a_2 - a_1} \lambda(\omega)[\theta(\omega) - a_2]^2 \quad (1)$$

and hence, for all $a \in [a_1, a_2]$, $T_z(a) < \infty$, since

$$T_z(a) \leq \frac{a_2 - a}{a_2 - a_1} T_z(a_1) + \frac{a - a_1}{a_2 - a_1} T_z(a_2) \quad (2)$$

Now let $b \in [a_1, a_2]$, and let c be any arbitrary point in A with $c \neq b$. Then from the identity

$$(b - \theta)^2 = (b - c + c - \theta)^2$$
$$= (b - c)^2 + 2(b - c)(c - \theta) + (c - \theta)^2 \quad (3)$$

we readily obtain

$$(b - c)^2 + 2(b - c)(c - \theta) \leq (b - \theta)^2 \quad (4)$$

and consequently we conclude that

$$E_\xi\{[\lambda(b - c)^2 + 2\lambda(b - c)(c - \theta)]|z\} < \infty \quad (5)$$

Setting $c = 0$, we get

$$E_\xi[(b\lambda - 2\theta\lambda)|z] < \infty \quad (6)$$

Since b is any arbitrary point in $[a_1, a_2]$, it follows that $E_\xi(\lambda|z) < \infty$ and therefore $E_\xi(\theta\lambda|z) < \infty$. The finiteness of $E_\xi(\lambda\theta^2|z)$ now follows from the finiteness of $T_z(a_1) < \infty$. Thus $T_z(a)$ is finite for all $a \in A$.

We next show that if $T_z(a) < \infty$ and $E_\xi(\lambda|z) < \infty$, then $T_z(a) < \infty$ for all $a \in A$.

Since $\lambda(\omega) > 0$, Schwarz' inequality yields

$$\begin{aligned}|E_\xi[\lambda(\theta - a_1)|z]| &\le E_\xi(\lambda|\theta - a_1| \,|z) = E_\xi[\sqrt{\lambda}\,(\sqrt{\lambda}|\theta - a_1|)|z]\\ &\le [E_\xi(\lambda|z)]^{\frac{1}{2}}[E_\xi(\lambda|\theta - a_1|^2|z)]^{\frac{1}{2}}\\ &= [E_\xi(\lambda|z)]^{\frac{1}{2}}[T_z(a)]^{\frac{1}{2}} < \infty\end{aligned} \qquad (7)$$

But, since

$$E_\xi(\lambda\theta|z) = E_\xi[a_1\lambda + \lambda(\theta - a_1)|z] \qquad (8)$$

we have by (7) and the assumption of finiteness of $E_\xi(\lambda|z)$ that $E_\xi(\lambda\theta|z) < \infty$. The finiteness of $E_\xi(\lambda\theta^2|z)$ follows from the finiteness of $T_z(a_1)$. Thus $T_z(a) < \infty$ for all $a \in A$.

Finally, if $T_z(a) < \infty$ for all $a \in A$, then $E_\xi(\lambda|z) < \infty$. This follows since in that case (6) holds for all $b \in A$.

From the theorem we see that if $T_z(a)$ exists for two values of $a \in A$, or, equivalently, if $T_z(a)$ exists for one value of a and $E_\xi(\lambda|z) < \infty$, then $T_z(a)$ exists for all a, and, moreover,

$$E_\xi(\lambda\theta^i|z) < \infty \qquad i = 0, 1, 2 \qquad (9)$$

Hence $T_z(a)$ can be written as

$$T_z(a) = a^2 E_\xi(\lambda|z) - 2aE_\xi(\lambda\theta|z) + E_\xi(\lambda\theta^2|z) \qquad (10)$$

which for a given z is a quadratic function in a and therefore has a minimum at $a^* = t(z)$, where

$$t(z) = \frac{E_\xi(\lambda\theta|z)}{E_\xi(\lambda|z)} \qquad (11)$$

We thus have the following.

Theorem 2

In any estimation problem where the loss is defined by

$$L = \lambda(\omega)(\theta - a)^2$$

if the a priori distribution ξ admits any estimate of finite expected risk, then it admits a Bayes estimate, and essentially only one, defined thus:

$$t^*(z) = \begin{cases} b & \text{if } T_z(b) < \infty \quad \text{for one } b \in A \\ \dfrac{E_\xi(\lambda\theta|z)}{E_\xi(\lambda|z)} & \text{if } T_z(a) < \infty \quad \text{for all } a \in A \end{cases}$$

Theorem 3

Let the loss function be given by $L = \lambda(\omega)(\theta - a)^2$. Then if for a given a priori distribution ξ which admits an estimate of finite risk the expectation of the function λ exists, the Bayes estimate t^* is biased, (i.e., the expectation of t^* is not identically equal to θ) or the Bayes risk is zero.

Proof:

Let $E_\xi(\lambda)$ be the expected value of λ with respect to an a priori distribution ξ. Then the existence of $E_\xi(\lambda)$ implies the existence of $E_\xi(\lambda|z)$ for all z for which $P_\xi(z) > 0$, as can be seen from the equation

$$E_\xi(\lambda) = \sum_z E_\xi(\lambda|z) P_\xi(z) \tag{12}$$

We define

$$\xi_\lambda(\omega) = \frac{\lambda(\omega)\xi(\omega)}{E_\xi(\lambda)} \tag{13}$$

Then the Bayes estimate with respect to the loss function L and ξ is the same as the Bayes estimate with respect to the loss function L^* defined by

$$L^*(\omega, a) = [\theta(\omega) - a]^2 \tag{14}$$

and the a priori distribution ξ_λ. Since ξ, and hence ξ_λ, admits an estimate of finite risk, the Bayes estimate with respect to ξ_λ exists and is given by

$$t^*(z) = E_{\xi_\lambda}(\theta|z) = E_{\xi_\lambda}(\theta|t^*) \tag{15}$$

Now suppose t^* is unbiased, that is,

$$E(t^*|\omega) = \theta(\omega) \tag{16}$$

for all ω. Then

$$E(\theta t^*) = E[t^* E(\theta|z)] = E(t^{*2}) = E[\theta E(t^*|\omega)] = E(\theta^2) \tag{17}$$

from which it follows that

$$E(L(\xi, t^*)) = \sum_\omega \sum_\xi [\theta(\omega) - t^*(z)]^2 p(z|\omega) \xi(\omega) = 0 \tag{18}$$

The assumption of the existence of the second moments of θ and t^*, which is implied in the above proof, can be removed.

If $E_\xi(\lambda)$ does not exist, the theorem no longer holds, as can be seen from the following example.

EXAMPLE

It is desired to obtain a Bayes estimate of a binomial probability ω of a success, on the basis of N independent trials when the a priori distribution ξ is rectangular over the interval $(0,1)$ and the loss function L is given by

$$L(\omega,a) = \frac{(\omega - a)^2}{\omega(1 - \omega)} \qquad 0 \leq a \leq 1$$

Here

$$\lambda(\omega) = \frac{1}{\omega(1 - \omega)} = \frac{1}{\omega} + \frac{1}{1 - \omega}$$

so that $\int_0^1 \lambda(\omega)\, d\omega = \infty$.

On the other hand, if m represents the number of successes in the N trials, then by the previous theorem the Bayes estimate $t^*(m)$ if $m \neq 0$ or 1 is given by

$$t^*(m) = \frac{\int_0^1 \frac{\omega^{m+1}(1 - \omega)^{N-m}}{\omega(1 - \omega)}\, d\omega}{\int_0^1 \frac{\omega^m(1 - \omega)^{N-m}}{\omega(1 - \omega)}\, d\omega} = \frac{m}{N}$$

which is unbiased.

7 □ BAYES ESTIMATES WHERE THE LOSS DEPENDS ON THE ABSOLUTE ERROR

As in the case of the quadratic risk function, let $T_z(a)$ be the conditional expected value of the loss function

$$L(\omega,a) = \lambda(\omega)|\theta(\omega) - a| \qquad \lambda(\omega) > 0$$

with respect to a fixed a priori distribution ξ when z is observed. Then

$$T_z(a) = E_\xi[\lambda(\omega)|\theta(\omega) - a|\,|z]$$

Theorem

For each z, $T_z(a) < \infty$ either for no a, for exactly one a, or for every $a \in A$. The second case implies $E_\xi(\lambda|z) = \infty$, and the third case implies $E_\xi(\lambda|z) < \infty$.

We first show that if $T_z(a) < \infty$ for a_1 and a_2 with $a_1 < a_2$, then $E_\xi(\lambda|z) < \infty$. This follows since

$$\lambda(\omega)(a_2 - a_1) \leq \lambda(\omega)[|a_2 - \theta(\omega)| + |\theta(\omega) - a_1|] \tag{1}$$

and hence

$$(a_2 - a_1)E_\xi(\lambda|z) \leq T_z(a_2) + T_z(a_1) < \infty \tag{2}$$

Now consider any $b \in A$, and let $a = a_1 - b$. Then

$$\lambda(\omega)|\theta(\omega) - a| \leq \lambda(\omega)|\theta(\omega) - a_1| + \lambda(\omega)|b| \tag{3}$$

and hence, since, by (2), $E_\xi(\lambda|z) < \infty$, we have

$$T_z(a) \leq T_z(a_1) + |b|E_\xi(\lambda|z) < \infty \tag{4}$$

and since b is arbitrary, we see that $T_z(a) < \infty$ for all $a \in A$, which proves the theorem.

8 □ RELATION OF THE PRINCIPLES OF INVARIANCE, MINIMAXITY, AND ADMISSIBILITY

These relations are demonstrated by assuming the following setup and proving the theorems.

Let the sample space be $(\mathfrak{X}, \mathfrak{A}, \{f_\lambda | \lambda \in \mathfrak{A}\})$. $r(\lambda)$ is a function of λ, $\mathfrak{D}$ is the set of estimates, and $\mathfrak{J}$ is a σ-field of subsets of $\mathfrak{D}$.

If $d \in \mathfrak{D}$, then $\{d\} \in \mathfrak{J}$.

$L(d,\lambda)$ is a loss function.

$m(\ |x)$, m is a probability measure over $(\mathfrak{D}, \mathfrak{J})$. G is the group of transformations such that $p_\lambda g^{-1} = p_{\bar{g}\lambda}$. $\bar{G}$ is the group of transformations such that $\bar{g}(r\lambda) = r(\bar{g}\lambda)$. $\bar{\bar{G}}$ is the group of transformations such that $\omega(\bar{\bar{g}}d, \bar{g}\lambda) = L(d, \lambda)$. m is invariant if

$$m(F|x) = m(\bar{\bar{g}}F|gX)$$
$$F \in \mathfrak{J} \quad g \in G \quad x \in \mathfrak{X}$$
$$\bar{\bar{g}}F = \{\bar{\bar{g}}d | d \in F\}$$

Suppose that m satisfies the following for $x \in \mathfrak{X}$: there is a

$$d \in \mathfrak{D} \rightarrow \{m(d)|x\} = 1.$$

Let $\delta(x)$ satisfy $\{m[\delta(x)]|x\} = 1$

$$m(F|x) = \begin{cases} 1 & \text{if } \delta(x) \in F \\ 0 & \text{if } \delta(x) \notin F \end{cases}$$

$$m(\bar{g}F|gx) = 1 \Leftrightarrow \delta(gx) \in \bar{g}F \Leftrightarrow \bar{g}^{-1}\delta(gx) \in F$$

If δ is invariant,

$$g^{-1}\delta(gx) = \delta(x) \Leftrightarrow \delta(x) \in F \Leftrightarrow m(F|x) = 1$$

Therefore $m[\bar{g}F|g(x)] = m(F|x) = 1$. m is invariant; therefore F is.

Theorem 1

Suppose m is any randomized estimate. For each $g \in G$ let mg be the randomized estimate such that

$$mg(F) = m[\bar{g}F|g(x)] \quad \begin{array}{l} F \in \mathfrak{F} \\ x \in \mathfrak{X} \end{array}$$

Then

$$R_m(\bar{g}\lambda) = R_{mg}(\lambda) \quad \begin{array}{l} g \in G \\ \lambda \in \Lambda \end{array}$$

Note that $mg = m$ if m is invariant.

$$mg^{-1}(F|gx) = m(\bar{g}^{-1}F|x)$$

$$R_m(\lambda) = \int_{\mathfrak{X}} \int_{\mathfrak{D}} L(d,\theta) \, dm(d|x) \, dp_\lambda(x)$$

$$= \int_{\mathfrak{X}} \int_{\mathfrak{D}} L(\bar{g}d, \bar{g}\lambda) \, dm(d|x) \, dp_\lambda(x)$$

$$= \int_{\mathfrak{X}} \int_{\mathfrak{D}} L(d, \bar{g}\lambda) \, dm_g^{-1}(d|gx) \, dp_\lambda(x)$$

$$= \int_{\mathfrak{X}} \int_{\mathfrak{D}} L(d, \bar{g}\lambda) \, dm_g^{-1}(d|x) \, dp_{\bar{g}\lambda}$$

$$= R_{mg}^{-1}(\bar{g}\lambda)$$

$$R_{mg}(\bar{g}^{-1}\lambda) = R_m(\lambda)$$

$$R_{mg}(\lambda) = R_m(\bar{g}\lambda)$$

Theorem 2

Let G be finite. If m is an estimate, then there is an invariant estimate m^* such that

$$\sup_{\lambda \in \Lambda} R_{m^*}(\lambda) \leq \sup_{\lambda \in \Lambda} R_m(\lambda)$$

Thus if a minimax estimate exists, an invariant minimax estimate exists.

Proof:

Let N be the number of elements in G.
$$m_g(F|x) = m(\bar{g}F|gx) \qquad g \in G$$
$$m^*(F|x) = \frac{1}{N} \sum_{g \in G} mg(F|x)$$

Let $g \in G$.
$$m^*(F|x) = \frac{1}{N} \sum_{h \in G} m_h(F|x)$$
$$= \frac{1}{N} \sum_{h \in G} m(\bar{h}F|hx)$$
$$= \frac{1}{N} \sum_{h \in G} m(\bar{h}g^{-1}\bar{g}F|hg^{-1}gx)$$
$$= \frac{1}{N} \sum_{h \in G} m_{hg^{-1}}(\bar{g}F|gx)$$
$$= \frac{1}{N} \sum_{k \in G} m_k(\bar{g}F|gx)$$
$$= m^*(\bar{g}F|gx)$$
$$hg^{-1} \neq h^*g^{-1} \qquad \text{if } h \neq h^*$$

Therefore
$$hg^{-1}g \neq h^*g^{-1}g$$

There is no repetition; thus m^* is an invariant.

$$R_{m^*}(\lambda) = \int_{\mathfrak{X}} \int_{\mathfrak{D}} L(d,\lambda)\, dm^*(d|x)\, dp_\lambda(x)$$
$$= \int_{\mathfrak{X}} \int_{\mathfrak{D}} L(d,\lambda)\, d\left[\frac{1}{N} \sum_{g \in G} mg(d|x)\right] dp_\lambda(x)$$
$$= \frac{1}{N} \sum_{g \in G} r_{mg}(\lambda)$$

m^* is invariant.
$$\sup_{\lambda \in \Lambda} R_{m^*}(\lambda) \leq \sup_{\lambda \in \Lambda} R_m(\lambda)$$
$$R_{m^*}(\lambda) = \frac{1}{N} \sum_{g \in G} R_{mg}(\lambda)$$
$$\sup_{\theta \in \Omega} R_{m^*}(\lambda) \leq \frac{1}{N} \sum_{g \in G} \sup_{\theta \in \Omega} R_{mg}(\lambda)$$
$$= \frac{1}{N} \sum_{g \in G} \sup_{\lambda \in \Lambda} R_m(\bar{g}\lambda)$$
$$= \sup_{\lambda \in \Lambda} R_m(\lambda)$$

Theorem 3

G is finite. Let m^{**} be an invariant estimate which is not admissible; then there is an invariant estimate m^* which is better than m^{**}.

Proof:

There is m such that

$$R_m(\lambda) \leq R_{m^{**}}(\lambda) \quad \text{for all } \lambda$$

with strict inequality for some λ.

$$R_{m^*}(\lambda) = \frac{1}{N} \sum_{g \in G} R_{mg}(\lambda)$$

$$= \frac{1}{N} \sum_{g \in G} R_m(\bar{g}\lambda)$$

$$\leq \frac{1}{N} \sum_{g \in G} R_{m^{**}}(\bar{g}\lambda)$$

$$= R_{m^{**}}(\lambda)$$

Since m^{**} is invariant.

Corollary

G is finite. If m is an invariant estimate such that there is no invariant estimate better than m, then m is admissible.

Definition

Let $x_1, \ldots, x_n$ be independent random variables with common density function. δ is invariant if

$$\delta(x_1 + b, \ldots, x_n + b) = \delta(x_1, \ldots, x_n) + b,$$

where b is any constant.

Theorem 4

If each invariant estimate has infinite risk, then each is uniformly minimum-risk (UMR) invariant. If at least one invariant estimate is with finite risk, then

$$x_1 - E_\theta(x_1 | x_2 - x_1, \ldots, x_n - x_1)$$

is UMR invariant.

ADMISSIBLE AND MINIMAX ESTIMATION 211

Proof:

Let δ have finite risk.

$$\begin{aligned}
R_\delta(0) &= \beta E_0[\delta(x_1, \ldots, x_n)]^2 \\
&= \beta E_0(E\{[\delta(0, x_2 - x_1, \ldots, x_n - x_1) \\
&\qquad\qquad + x_1]^2 | x_2 - x_1, \ldots, x_n - x_1\}) \\
&\geq \beta E_0(E_0\{[-E(x_1|x_2 - x_1, \ldots, x_n - x_1) \\
&\qquad\qquad + x_1]^2 | x_2 - x_1, \ldots, x_n - x_1\}) \\
&= R_{\delta_0}(0)
\end{aligned}$$

The conditional probability density is

$$p(x_1 | x_2 - x_1, \ldots, x_n - x_1)\Big|_{y_2, \ldots, y_n}$$

$$= \begin{cases} \dfrac{p(x_1, x_1 + y_2, \ldots, x_1 + y_n)}{q(y_2, \ldots, y_n)} & \begin{array}{l}\text{if } x_2 - x_1 = y_2 \\ \phantom{\text{if }} x_3 - x_1 = y_3 \\ \phantom{\text{if }}\cdots\cdots\cdots \\ \phantom{\text{if }} x_n - x_1 = y_n\end{array} \\ 0 & \text{otherwise} \end{cases}$$

$$q(y_2, \ldots, y_n) = \int_{-\infty}^{\infty} p(x_1, x_1 + y_2, \ldots, x_1 + y_n)\, dx_1$$

$$E_0(x_1 | x_2 - x_1, \ldots, x_n - x_1) = \frac{\int_{-\infty}^{\infty} x_1 p(x_1, x_1 + y_2, \ldots, x_1 + y_n)\, dx_1}{\int_{-\infty}^{\infty} p(x_1, x_1 + y_2, \ldots, x_1 + y_n)\, dx_1}$$

EXAMPLE

Let $X_1, \ldots, X_n$ be independent and each $N(\lambda, 1)$. For $\lambda = 0$

$$p(x_1, \ldots, x_n) = \frac{1}{(\sqrt{2\pi})^n} e^{-\Sigma x_i^2/2}$$

$$E_0(x_1 | x_2 - x_1, \ldots, x_n - x_1) + \sum_{i=1}^{n} \frac{y_i}{n}$$

$$= \frac{\dfrac{1}{n}\int_{-\infty}^{\infty} \left(nx_1 + \sum_{i=1}^{n} y_i\right) e^{-\frac{1}{2}\Sigma(x_1+y_i)^2}\, dx_1}{\int_{-\infty}^{\infty} e^{-\frac{1}{2}\Sigma(x_1+y_i)^2}\, dx_1} = 0$$

where

$$y_1 = 0 \qquad y_i = x_i - x_1$$

$$E_0(x_1 | x_2 - x_1, \ldots, x_n - x_1) + \sum_{i=1}^{n} \frac{y_i}{n} = 0$$

$$E(x_1 | x_2 - x_1, \ldots, x_n - x_1) = -\bar{x} + x_1$$

Therefore
$$x_1 - E(x_1|x_2 - x_1, \ldots, x_n - x_1) = \bar{x}$$
and $\bar{X}$ is the UMV invariant estimate of $N(\lambda,1)$

PROBLEMS

1. (Lehmann and Hodges) Let X be a random variable with hypergeometric density function
$$P(X = k) = \frac{\binom{D}{k}\binom{n-D}{n-k}}{\binom{N}{n}}$$
where k, D, n and N are positive integers such that $k < D$, $K < n$, $n < N$, and $D < n$, and let $(\delta - D)^2$ be the loss function. Prove that
$$\frac{N}{n + \sqrt{n(N-n)/(N-1)}} X + \frac{N}{2}\left[1 - \frac{n}{n + \sqrt{(N-n)n/(N-1)}}\right]$$
is the minimax estimate of D. *Hint:* use
$$P(D = d) = \int_0^1 \binom{N}{d} p^d q^{N-d} C p^{a-1} q^{b-1}\, dp$$
as the a priori distribution, where $a, b > 0$ and $C = \Gamma(a+b)/[\Gamma(a)\Gamma(b)]$

2. (Wasan) Let X be a random variable with geometric density function
$$P(X = k) = pq^{k-1} \quad \begin{array}{l} k = 1, 2, \ldots, \infty \\ 0 < p \leq 1 \\ q = 1 - p \end{array}$$
and let $(\delta - p)^2$ be the loss function. Then prove that
$$p = \begin{cases} \frac{3}{4} & \text{when } k = 1 \\ \frac{1}{4} & \text{when } k \geq 2 \end{cases}$$
is a minimax estimate of p. *Hint:* Use the following a priori distribution of p:
$$p = \begin{cases} \frac{1}{4} & \text{with probability } 2/3 \\ 1 & \text{with probability } 1/3 \end{cases}$$

3. (Wasan) Let X be a random variable with density function

$$p(X = k) = qp^{k-1} \qquad k = 1, 2, \ldots, \infty$$
$$0 \le p < 1$$

For a squared-error loss function $(\delta - p)^2$, find the minimax estimate of p.

4. (Katz) Let X be a random variable with exponential density function

$$P_\omega(x) = \beta(\omega)e^{x\omega} \qquad \text{for all } x$$
$$\omega \in T$$

where

$$T = \{\omega | \beta(\omega)^{-1} = \int_x e^{x\omega}\, d\mu(x) < \infty\}$$

μ is a σ-finite measure on the real line with a nondegenerate spectrum. Let $\Omega = \{\omega | \omega \ge a\} \subset T$, where a is an interior point of T. Taking the squared error as the loss function, prove that

$$\delta(X) = X + \frac{\beta(a)e^{aX}}{\int_a^\infty \beta(\omega)e^{\omega X}\, d\omega}$$

is an admissible estimate of $E_\omega(X)$.

In particular show the following:

(a) Let X be a binomial random variable with parameter p where $\Omega = \{p | p \ge a\}$; then

$$\delta(X) = X + \frac{a^X(1-a)^{n-X}}{\int_a^1 p^{X-1}(1-p)^{n-X-1}\, dp} \qquad x = 0, 1, 2, \ldots, n$$

(b) Let X be a Poisson random variable with parameter λ, where $\Omega = \{\lambda | \lambda \ge a\}$; then

$$\delta(X) = X + \frac{a^X e^{-a}}{\int_a^\infty \lambda^{X-1} e^{-a}\, dX} \qquad x = 0, 1, 2, \ldots, \infty$$

(c) Let X be a gaussian random variable with parameter ω and $\Omega = \{\omega | \omega \ge 0\}$; then

$$\delta(X) = X + \frac{e^{-X^2/2}}{\int_{-\infty}^X e^{-t^2/2}\, dt}$$

5. (Karlin) Let X be a random variable with exponential density as defined above. Let

$$\int_c^b \beta^{-\lambda}(\omega)\, d\omega \to +\infty \qquad \text{as } b \to \bar{\omega}$$

$$\int_a^c \beta^{-\lambda}(\omega)\, d\omega \to +\infty \qquad \text{as } a \to \underline{\omega}$$

where c is an interior point of $\Omega = (\underline{\omega},\bar{\omega})$. Prove that $X/(\lambda + 1)$ is an admissible estimate of $E_\omega(X)$, where $\underline{\omega} \to -\infty$ and $\bar{\omega} \to +\infty$.

6. Let X be a random variable with density function

$$P(x = 1) = p \qquad 0 \le p \le 1$$
$$P(x = 0) = 1 - p = q$$

Let $\delta(X)$ be an estimate of p, and let the loss function be

$$L[p,\delta(X)] = \begin{cases} 1 & \text{if } |\delta(X) - p| \ge \tfrac{1}{4} \\ 0 & \text{if } |\delta(X) - p| < \tfrac{1}{4} \end{cases}$$

Find a minimax estimate of p. Show that for every nonrandomized estimate $\delta(X)$, $\sup_p E[L(p,\delta)] = 1$. Construct a randomized minimax estimate δ^* and compare $\sup_p E\{L[p,\delta(X)]\}$ and $\sup_p E[L(p,\delta^*)]$.

7. Let X be a random variable with density function

$$P(X = k) = \frac{\lambda^k e^{-\lambda}}{k!} \qquad k = 0, 1, 2, \ldots, \infty$$

Let $[\delta(X) - \lambda]^2/\lambda$ be the loss function. By the method of the Cramér-Rao inequality, find a minimax estimate of λ.

8. Let X be a random variable with inverse gamma density function with parameter α. Construct a minimax estimate of α for a properly scaled quadratic loss function which gives constant risk.

9. (a) Let X be a random variable with the following density functions. Compare the risk of the minimax estimate, the uniformly minimum-variance estimate, and the maximum-likelihood estimate of the parameter involved taking the squared error as the loss function.

(i) $P(X = k) = \binom{n}{k} p^k q^{n-k} \qquad k = 0, 1, 2, \ldots, n$
$$0 \le p \le 1$$
$$q = 1 - p$$

(ii) $P(X = k) = pq^{k-1} \quad k = 1, 2, \ldots, \infty$
$0 < p \leq 1$

(iii) $P_\mu(x) = \dfrac{1}{\sqrt{2\pi}} e^{-(x-\mu)^2/2} \quad -\infty < x < \infty$
$-\infty < \mu < \infty$

(b) Prove that when a minimax estimate of the parameter is unique for the squared-error loss function, then it is admissible.

10. Obtain the minimax invariant estimate of λ based on n independent observations from a rectangular distribution defined as

$$f_\lambda(x) = \begin{cases} 1 & \lambda < x < \lambda + 1 \\ & \lambda > 0 \\ 0 & \text{otherwise} \end{cases}$$

and take the squared error as the loss function.

11. If $X_1, X_2, \ldots, X_n$ are independent random variables with density function

$$P_\lambda(X_i = j) = \frac{1}{\lambda} \quad \begin{matrix} j = 1, 2, \ldots, \lambda \\ \lambda = 1, 2, \ldots \end{matrix}$$

and the loss function is $L(\delta\lambda) = (\delta - \lambda)^2$, show that

$$\delta(X_1, \ldots, X_n) = \frac{2(X_1 + X_2 + \cdots + X_n)}{n} - 1$$

is not an admissible estimate for λ. Also find a uniformly minimum-variance-unbiased estimate of λ.

12. Let $\delta_i(x)$ be a constant-risk estimate of λ_i, and suppose that for any other estimate $\bar{\delta}_i(x)$ there exists a probability measure $\mu(\lambda)$ over Ω such that

$$\int_\Omega r_{\delta_i}(\lambda) \, d\mu(\lambda) \leq \int_\Omega r_{\bar{\delta}_i}(\lambda) \, d\mu(\lambda)$$

Then prove that $\delta_i(x)$ is a minimax estimate of λ_i, where $r = E[L(\delta, \lambda)]$, L being a loss function.

13. If η is a complete family of estimates and A is the admissible family of estimates, prove that A is a subset of η.

14. If ξ is an essentially complete family of estimates and there exists an admissible estimate δ_0 not in ξ, prove that there exists a δ in ξ which is equivalent to δ_0.

15. Let $X_1, X_2, \ldots, X_n$ be independent random variables with common density function

$$f(x,\mu,\lambda) = \begin{cases} \dfrac{\sqrt{\lambda}}{\sqrt{2\Pi x^3}} \exp\left(-\dfrac{\lambda(x,\mu)^2}{2\mu^2 x}\right) & \begin{array}{l} x > 0 \\ \lambda > 0 \\ \mu > 0 \end{array} \\ 0 & \text{otherwise} \end{cases}$$

Prove that δ is an admissible and minimax estimate of $1/\lambda$ for loss function $\lambda^2(\delta - 1/\lambda)^2$, where

$$\delta = \frac{1}{n+1} \sum_{i=1}^{n} \left(\frac{1}{X_i} - \frac{1}{\bar{X}}\right)$$

REFERENCES

1 BLACKWELL, D., and M. A. GIRSCHICK: "Theory of Games and Statistical Decision Functions," John Wiley & Sons, Inc., New York, 1954.

2 BLYTH, C. R.: On Minimax Statistical Decision Procedures and Their Admissibility, *Ann. Math. Statist.*, vol. 22, pp. 22–42, 1951.

3 GIRSCHICK, M. A., and L. J. SAVAGE: Bayes and Minimax Estimates for Quadratic Loss Function, *Proc. 2d Berkeley Symp. Math. Statist. Probability*, pp. 53–73, 1951.

4 HODGES, J. L., JR., and E. L. LEHMANN: Some Applications of the Cramér-Rao Inequality, *Proc. 2d Berkeley Symp. Math. Statist. Probability*, pp. 13–22, 1951.

5 HODGES, J. L., JR., and E. L. LEHMANN: Some Problems in Minimax Point Estimation, *Ann. Math. Statist.*, vol. 21, pp. 182–197, 1950.

6 KARLIN, S.: Admissibility for Estimation with Quadratic Loss, *Ann. Math. Statist.*, vol. 29, pp. 406–436, 1958.

7 KATZ, M. W.: Admissible and Minimax Estimates for Parameters in Truncated Spaces, *Ann. Math. Statist.*, vol. 32, pp. 136–142, 1961.

8 KUDO, H.: On Minimax Invariant Estimates of the Transformation Parameter, *Nat. Sci. Rept. Ochanomizu Univ. Tokyo*, pp. 31–73, 1955.

9 WALD, A.: Statistical Decision Functions Which Minimize the Maximum Risk, *Ann. Math.*, vol. 46, p. 265, 1945.

10 WASAN, M. T.: Minimax Estimate of a Negative Binomial Parameter, abstract, *Ann. Math. Statist.*, vol. 33, p. 1501, 1962.

11 WASAN, M. T.: Sequential Estimation of a Binomial Parameter, *Proc. Intern. Symp. Classical Contagious Discrete Distributions, Montreal*, pp. 263–272, 1963.

12 WOLFOWITZ, J.: Minimax Estimate of the Mean of a Normal Distribution with Known Variance, *Ann. Math. Statist.*, vol. 21, pp. 218–230, 1950.

chapter Eleven

The Empirical Bayes Method of Estimation

1 □ INTRODUCTION

In his pioneer paper, Robbins [28] gives an empirical approach to a problem of decision and, in particular, to a problem of estimation. In fact, he takes a middle road, between the bayesian and nonbayesian approaches. For a bayesian, a parameter is a random variable with known density function, which is the point of view we took in the previous chapter in order to determine admissible and minimax estimates. Robbins [28, 30] assumes that it exists but is not known.

In this chapter we give an empirical Bayes technique and its optimality criterion, prove a couple of theorems, and state others without proof but illustrate them by examples. Then an example is given where an asymptotically optimal procedure does not exist, but a reasonable approximation to one can be found. We also append a method of Maritz [16]. We attempt to illustrate theoretical results as well as examples for a squared-error loss function, though these may be true for general loss functions. The theory involves estimation of the density function, which is beyond the scope of this volume. However, we append a number of problems for solution which also reveal the character of this procedure.

2 □ AN EMPIRICAL BAYES TECHNIQUE

Let X be a random variable which has a discrete probability distribution depending in a known way on an unknown real parameter Λ,

$$P(x|\lambda) = P_r(X = x | \Lambda = \lambda) \tag{1}$$

Λ itself being a random variable with a priori distribution function

$$G(\lambda) = P_r(\Lambda \leq \lambda) \tag{2}$$

The marginal probability distribution of X is then given by

$$P_G(x) = P_r(X = x) = \int P(x|\lambda)\, dG(\lambda) \tag{3}$$

and the expected squared loss or risk of any estimator of λ of the form $\phi(x)$ is

$$\begin{aligned}
E[\phi(x) - \Lambda]^2 &= E(E\{[\phi(x) - \Lambda]^2 | \Lambda = \lambda\}) \\
&= \int \sum_x p(x|\lambda)[\phi(x) - \lambda]^2\, dG(\lambda) \\
&= \sum_x \int p(x|\lambda)[\phi(x) - \lambda]^2\, dG(\lambda) \tag{4}
\end{aligned}$$

which is a minimum when $\phi(x)$ is defined for each x as that $y = y(x)$ for which

$$I(x) = \int p(x|\lambda)(y - \lambda)^2\, dG(\lambda) = \min \tag{5}$$

But for any fixed x the quantity

$$\begin{aligned}
I(x) &= y^2 \int p\, dG - 2y \int p\lambda\, dG + \int p\lambda^2\, dG \\
&= \int p\, dG \left(y - \frac{\int p\lambda\, dG}{\int p\, dG} \right)^2 + \left[\int p\lambda^2\, dG - \frac{(\int p\lambda\, dG)^2}{\int p\, dG} \right] \tag{6}
\end{aligned}$$

is a minimum with respect to y when

$$E(\lambda|x) = y = \frac{\int p\lambda\, dG}{\int p\, dG} \tag{7}$$

the minimum value of $I(x)$ being

$$I_G(x) = \int p(x|\lambda)\lambda^2 \, dG(\lambda) - \frac{[\int p(x|\lambda)\lambda \, dG(\lambda)]^2}{\int p(x|\lambda) \, dG(\lambda)} \qquad (8)$$

Hence the Bayes estimate of Λ corresponding to the a priori distribution function G of Λ is the random variable $\phi_G(x)$ defined by the function

$$T = \phi_G(x) = \frac{\int p(x|\lambda)\lambda \, dG(\lambda)}{\int p(x|\lambda) \, dG(\lambda)} \qquad (9)$$

the corresponding minimum value of (4) being

$$E[\phi_G(x) - \Lambda]^2 = \sum_x I_G(x) \qquad (10)$$

Expression (9) is, of course, the expected value of the a posteriori distribution of Λ given $X = x$. Generally G is not known, and thus we cannot compute $\phi_G(x)$. We discuss a technique for estimating G which is due to Robbins [28].

Suppose now that the problem of estimating the realization λ from an observed value of X is going to occur repeatedly with a fixed and known $p(x|\lambda)$ and a fixed but unknown $G(\lambda)$, and let

$$(\Lambda_1, X_1), (\Lambda_2, X_2), \ldots, (\Lambda_n, X_n) \qquad (11)$$

denote the sequence of random variables so generated. [The Λ_r are independent random variables with common distribution function G, and the distribution of X_n depends only on Λ_n; for $\Lambda_n = \lambda$ it is given by $p(x|\lambda)$.] If we wanted to estimate an unknown λ_n for an observed X_n, and if the previous values $\Lambda_1, \ldots, \Lambda_{n-1}$ were by now known, we could form the empirical distribution function of the random variable Λ,

$$G_{n-1}(\lambda) = \frac{\text{no. of terms } \Lambda_1, \ldots, \Lambda_{n-1} \text{ which are } \leq \lambda}{n-1} \qquad (12)$$

and take as our estimate of λ_n the quantity $\psi_n(x_n)$, where by definition

$$\psi_n(x) = \frac{\int p(x|\lambda)\lambda \, dG_{n-1}(\lambda)}{\int p(x|\lambda) \, dG_{n-1}(\lambda)} \qquad (13)$$

Since $\Lambda_1, \ldots, \Lambda_{n-1}$ will not be known, one has to estimate G from $x_1, x_2, \ldots, x_n$. We observe that for any fixed x the empirical frequency

$$p_n(x) = \frac{\text{no. of } x_1, \ldots, x_n \text{ equal to } x}{n} \qquad (14)$$

tends with probability 1 as $n \to \infty$ to determine $p_G(x)$ defined by (3), no matter what the a priori distribution function G. Thus from an approximate value (14) of the integral (3), where $p(x|\lambda)$ is a known kernel, we obtain an approximation to the known distribution function G or at least, in the present case, to the value of the Bayes function (9), which depends on G. The probability of doing this will depend on the nature of the kernel $p(x|\lambda)$ and on the class, say $\mathcal{G}$, to which the unknown G is assumed to belong. Let us consider some examples to illustrate this fact.

EXAMPLE 1
Let
$$p(x|\lambda) = e^{-\lambda}\frac{\lambda^x}{x!} \qquad \begin{array}{l} x = 0, 1 \ldots \\ \lambda > 0 \end{array} \qquad (15)$$

and let $\mathcal{G}$ be the class of all distribution functions on the positive axis. We have
$$p_G(x) = \int p(x|\lambda) \, dG(\lambda)$$
$$= \frac{\int_0^\infty e^{-\lambda}\lambda^x \, dG(\lambda)}{x!} \qquad (16)$$

and
$$\phi_G(x) = \frac{\int_0^\infty \frac{\lambda e^{-\lambda}}{x!}\lambda^x \, dG(\lambda)}{\int_0^\infty \frac{e^{-\lambda}\lambda^x}{x!} \, dG(\lambda)} \qquad (17)$$

We can write this in terms of p_G:
$$\phi_G(x) = (x+1)\frac{p_G(x+1)}{p_G(x)} \qquad (18)$$

If we now define the function
$$\phi_n(x) = (x+1)\frac{p_n(x+1)}{p_n(x)}$$
$$= (x+1)\frac{\text{no. of terms } x_1, \ldots x_n \text{ equal to } x+1}{\text{no. of terms } x_1, \ldots, x_n \text{ equal to } x} \qquad (19)$$

then no matter what the unknown G, we shall have for any fixed x
$$\phi_n(x) \to \phi_G(x) \text{ with probability 1 as } n \to \infty \qquad (20)$$

A question arises whether the risk of $\phi_n(x)$ tends to the Bayes risk as $n \to \infty$. We answer this in Sec. 3.

EXAMPLE 2
Let
$$p(x|\lambda) = (1 - \lambda)\lambda^x \qquad \begin{array}{c} x = 0, 1, \ldots \\ 0 < \lambda < 1 \end{array}$$

and
$$p_G(x) = \int_0^1 (1 - \lambda)\lambda^x \, dG(\lambda)$$
$$\phi_G(x) = \frac{\int_0^1 (1 - \lambda)\lambda^{x+1} \, dG(\lambda)}{\int_0^1 (1 - \lambda)\lambda^x \, dG(\lambda)}$$
$$= \frac{p_G(x + 1)}{p_G(x)}$$

In this case one should take

$$\phi_n(x) = \frac{\text{no. of terms } x_1, \ldots, x_n \text{ equal to } x + 1}{\text{no. of terms } x_1, \ldots, x_n \text{ equal to } x}$$

EXAMPLE 3
Let
$$p_r(x|\lambda) = \binom{r}{x} \lambda^x (1 - \lambda)^{r-x} \qquad \begin{array}{c} x = 0, 1, 2, \ldots, r, \\ 0 \leq \lambda \leq 1 \end{array}$$

where r is positive integer. $\mathcal{G}$ may be taken as the class of all distribution functions on the interval $(0,1)$. Then

$$p_{G,r}(x) = \int p_r(x|\lambda) \, dG(\lambda)$$
$$= \binom{r}{x} \int_0^1 \lambda^x (1 - \lambda)^{r-x} \, dG(\lambda)$$
$$\phi_{G,r}(x) = \frac{\int_0^1 \lambda^{x+1}(1 - \lambda)^{r-x} \, dG(\lambda)}{\int_0^1 \lambda^x (1 - \lambda)^{r-x} \, dG(\lambda)}$$

In order to write the fundamental relation

$$\phi_{G,r}(x) = \frac{x + 1}{r + 1} \frac{p_{G,r+1}(x + 1)}{p_{G,r}(x)} \qquad x = 0, 1, 2, \ldots, r$$

one can take

$$p_{n,r}(x) = \frac{\text{no. of terms } x_1, x_1, \ldots, x_n \text{ equal to } x}{n}$$

Then $p_{n,r}(x) \to p_{G,r}(x)$ with probability 1 as $n \to \infty$. Thus one can have

$$\phi_{n,r}(x) = \frac{x+1}{r} \frac{p_{n,r+1}(x+1)}{p_{n,r}(x)}$$

However, one may not be able to estimate $P_{G,r+1}(x+1)$ unless the observations are in a special form. Here is a case where there may be no empirical Bayes estimate. We discuss this again in Sec. 4.

EXAMPLE 4
Let the density function be of laplacian type

$$p(x|\lambda) = e^{\lambda x} f(x) h(\lambda)$$
$$p_G(x) = f(x) \int e^{\lambda x} h(\lambda) \, dG(\lambda)$$
$$\phi_G(x) p_G(x) = f(x) \int \lambda e^{\lambda x} h(\lambda) \, dG(\lambda)$$
$$= f(x) \frac{d}{dx} \frac{p_G(x)}{f(x)}$$

provided the differentiation under the integral sign is justified. Hence

$$\phi_G(x) = \frac{d}{dx} \log \frac{p_G(x)}{f(x)}$$

In the above fashion one can approximate $p_G(x)$ by $p_n(x)$. Thus much depends on the form of the density function $p(x|\lambda)$ and the class $\mathcal{G}$ of distribution functions G. An empirical procedure for continuous density functions can be developed in a similar way. We give some illustrative examples.

EXAMPLE 5
Let X have the first-passage time density of standard brownian motion, that is,

$$g(x|\lambda) = \begin{cases} \left(\frac{\lambda}{2\pi x^3}\right)^{\frac{1}{2}} e^{-\lambda/2x} & x > 0 \\ & \lambda > 0 \\ 0 & \text{otherwise} \end{cases}$$

We would like to construct a consistent sequence of estimates for $E(\Lambda|x)$, where

$$E(\Lambda|x) = \frac{\int_\Omega \lambda g(x|\lambda)\,dG(\lambda)}{g_G(x)}$$

where

$$g_G(x) = \int_\Omega g(x|\lambda)\,dG(\lambda)$$

Now

$$\log g(x;\lambda) = -\tfrac{1}{2}\log 2\pi - \tfrac{3}{2}\log x + \tfrac{1}{2}\log \lambda - \frac{\lambda}{2x}$$

Therefore

$$\frac{\partial \log g}{\partial x} = -\frac{3}{2x} + \frac{\lambda}{2x^2}$$

$$\lambda = 3x + 2x^2\,\frac{\partial \log g}{\partial x}$$

$$\sum(\Lambda|x) = \int_\Omega \frac{[3x + 2x^2(\partial \log g)/\partial x]g(x|\lambda)\,dG(\lambda)}{g_G(x)}$$

$$= 3x + 2x^2 \frac{\int_\Omega \frac{\partial \log g}{\partial x} g(x|\lambda)\,dG(\lambda)}{g_G(x)}$$

$$= 3x + 2x^2 \frac{\int_\Omega \frac{\partial g(x|\lambda)}{\partial x}\,dG(\lambda)}{g_G(x)}$$

$$= 3x + 2x^2 \frac{g'_G(x)}{g_G(x)}$$

We assume the change in order of integration and differentiation where

$$g'_G(x) = \frac{dg_G(x)}{dx}$$

Now we substitute consistent estimates $g_n(x;\lambda)$ and $g'_n(x;\lambda)$ of $g_G(x)$ and $g'_G(x)$, respectively. Hence

$$E_n(\lambda|x) = 3x + 2x^2 \frac{g'_n(x;\lambda)}{g_n(x;\lambda)}$$

is a consistent estimate of $E(\lambda|x)$. See Prob. 2 for construction of $g_n(x;\lambda)$ and $g'_n(x;\lambda)$ and see also Roy and Wasan [31].

EXAMPLE 6

Let X have the first-passage time density of brownian motion with positive drift, that is,

$$f(x|\lambda) = \begin{cases} \dfrac{\sqrt{\lambda}}{\sqrt{2\pi x^3}} e^{\dfrac{-\lambda(x-1)^2}{2x}} & x > 0 \\ & \lambda > 0 \\ 0 & \text{otherwise} \end{cases}$$

We would like to construct a consistent sequence of estimates for $E(\lambda|x)$, where

$$E(\Lambda|x) = \frac{\int_\Omega \lambda f(x|\lambda)\, dG(\lambda)}{\int_\Omega f(x|\lambda)\, dG(\lambda)} = \frac{\int \lambda f(x|\lambda)\, dG(\lambda)}{f_G(x)} \qquad (21)$$

Let

$$f_G(x) = \int_\Omega f(x|\lambda)\, dG(\lambda)$$

Now

$$\log f(x|\lambda) = -\tfrac{1}{2}\log 2\pi - \tfrac{3}{2}\log x + \tfrac{1}{2}\log \lambda - \frac{\lambda x}{2} + \lambda - \frac{\lambda}{2x}$$

$$\frac{\partial \log f}{\partial x} = -\frac{3}{2x} - \frac{\lambda}{2} + \frac{\lambda}{2x^2}$$

$$\lambda = \frac{3/2x + (\partial \log f)/\partial x}{1/2x^2 - \tfrac{1}{2}} \qquad (22)$$

Substituting (22) in (21), we obtain

$$E(\Lambda|x) = \frac{3x}{1-x^2} + \frac{2x^2}{1-x^2} \frac{\int_\Omega \dfrac{\partial f(x|\lambda)}{\partial x}\, dG(\lambda)}{f_G(x)}$$

$$= \frac{3x}{1-x^2} + \frac{2x^2}{1-x^2} \frac{f_G'(x)}{f_G(x)}$$

assuming that reversing integration and differentiation is valid and $x^2 \neq 1$.

Now we substitute consistent estimates $f_n(x)$ and $f_n'(x)$ of $f_G(x)$ and $f_G'(x)$, respectively (see Prob. 2)

$$f_n(x) = \frac{1}{2\pi n^{\frac{1}{3}}} \sum_{i=1}^{n} \left(\frac{\sin y_i}{y_i}\right)^2$$

where
$$y_i = \frac{x - x_i}{2n^{-\frac{1}{5}}}$$
and
$$f'_n(x) = n^{\frac{1}{5}}[f_n(x + n^{-\frac{1}{5}}) - f_n(x)]$$
Therefore
$$E_n(\lambda|x) = \frac{3x}{1-x} + \frac{2x^2}{1-x^2}\cdot\frac{f'_n(x)}{f_n(x)}$$

is a consistent sequence of estimates of $E(\Lambda|x)$. (See Roy and Wasan [32].) The subject matter of Secs. 3 and 4 is a result of Robbins [30].

3 □ OPTIMALITY CRITERION

We shall now assume that the random variable X has a continuous density function $f_X(x)$.

Let us define an *empirical* or adaptive estimation procedure to be a sequence $\delta = \{\delta_n\}$ of the function of the form (13) of Sec. 2 with value in the set R. For the given δ

$$R_n(\delta, G) = \int_{-\infty}^{\infty}\int_{-\infty}^{\infty}\cdots\int_{-\infty}^{\infty}[\delta_n(x, \ldots, x_n) - \lambda]^2 \prod_1^n f_\lambda(x_i)\, dx_i\, dG(\lambda)$$
$$\geq R(G) = \min_\delta R(\delta, G)$$

Definition 1

If
$$\lim_{n \to \infty} R_n(\delta, G) = R(G)$$

we say that δ is asymptotically optimal relative to G.

We shall now prove a main theorem and state others but illustrate them by examples, since the problem of an empirical Bayes method leads to a problem of nonparametric estimation. The following is an adaptation of the theorem of Robbins [30]. We assume the setup of Sec. 2.

Theorem 1

Let

$$L(\delta,\lambda) = (\delta - \lambda)^2 \quad \text{where } \delta \text{ is an estimate of } \lambda \quad (1)$$
$$L(\lambda) = \sup_{\delta \in R} (\delta - \lambda)^2 \leq \infty \quad \text{where } R \text{ is the real line} \quad (2)$$

and

$$\int_\Omega L(\lambda)\, dG(\lambda) < \infty \quad (3)$$
$$H(x) = \int_\Lambda L(\lambda) f(x|\lambda)\, dG(\lambda) < \infty \quad (4)$$
$$\phi_G(\delta,x) = \int_\Omega L(\delta,\lambda) f(x|\lambda)\, dG(\lambda) \quad (5)$$

Let $\delta_n(x) = \delta_n(x;x_1, \ldots, x_n)$ be an estimate defined in (13) of Sec. 2 for every x which is based upon the sequence (11) of Sec. 2. Let

$$\delta_G(x) = E(\lambda|x) \quad (6)$$

such that

$$P[\lim_{n \to \infty} \delta_n(x) = \delta_G(x)] = 1$$
$$\phi_G[\delta_n(x),x] \to \phi_G[\delta_G(x),x] \quad (7)$$

in probability (where in probability refers to the random variables $X_1, X_2, \ldots, X_n$). Then

$$\lim_{n \to \infty} R[\delta_n(X),G] = R[\delta_G(X)] \quad (8)$$

We say that $\delta_n(X)$ is asymptotically optimal relative to G.

Proof:

The risk of an estimate $\delta_n(X)$ is

$$R[\delta_n(X),G] = \int E\{\phi_G[\delta_n(x),x]\}\, dx \quad (9)$$

where E represents an expectation with respect to the n independent random variables $X_1, X_2, \ldots, X_n$ and not with respect to x, the most recent observable random variable. To show (7) we first show that

$$\lim_{n \to \infty} E\{\phi_G[\delta_n(x),x]\} = \phi_G[\delta_G(x),x] \quad (10)$$

and then that

$$\lim_{n \to \infty} \int E\{\phi_G[\delta_n(x),x]\}\, dx = \int \lim_{n \to \infty} E\{\phi_G[\delta_n(x),x]\}\, dx \quad (11)$$

Define the function

$$H(x) = \int L(\lambda) f(x|\lambda)\, dG(\lambda)$$

Since $\delta_n(x) \in A$, we have from the definition in hypothesis (3) that

$$H(x) \geq \int L[\delta_n(x),\lambda] f(x|\lambda) \, dG(\lambda)$$

Since the loss is positive,

$$H(x) \geq |\phi_G[\delta_n(x),x]| \qquad (12)$$

$H(x)$ is integrable since

$$\int L(\lambda) \, dG(\lambda) = \int L(\lambda)[\int f(x|\lambda) \, dx] \, dG(\lambda)$$
$$= \int\int L(\lambda) f(x|\lambda) \, dG(\lambda) \, dx$$

Therefore by condition (3)

$$\int H(x) \, dx < \infty$$

$H(x)$ is also almost everywhere finite.

We now apply the Lebesgue bounded-convergence theorem (Halmos [4, p. 110]) to show (10). By hypothesis (7) we have a sequence of integrable functions converging in probability to a limit; by (12) each element in the sequence is bounded by $H(x)$. Thus we have that

$$\lim_{n \to \infty} E\{|\phi_G[\delta_n(x),x] - \phi_G[\delta_G(x),x]|\} = 0 \qquad \text{a.e. } x$$

Since for any arbitrary integrable functions f and g

$$|E(f) - E(g)| \leq E(|f - g|)$$

we obtain

$$\lim_{n \to \infty} E\{\phi_G[\delta_n(x),x]\} = E\{\phi_G[\delta_G(x),x]\} \qquad \text{a.e. } x \qquad (13)$$

The expectation is independent of x; hence we obtain (10) from (13).

To show (11) we will apply the Lebesgue dominated-convergence theorem. From (12) we see that each element in the sequence $\{E\{\phi_G[\delta_n(x),x]\}\}$ is dominated by $H(x)$, a.e. x. From (10) we see that the sequence converges a.e. x to the function $\phi_G[\delta_G(x),x]$. The conditions of the convergence theorem are satisfied. This completes the proof.

Let $x_1, x_2, \ldots, x_n, \ldots$ be a sequence of independent random variables with F_G as their common distribution function and define

$$B_n(x) = B_n(x_1, \ldots, x_n, x)$$
$$= \frac{\text{no. of terms which are} \leq x}{n} \qquad (14)$$

For any two distribution functions F_1, F_2 define the distance

$$\rho(F_1, F_2) = \sup_x |F_1(x) - F_2(x)| \qquad (15)$$

and let ε_n be any sequence of positive constants tending to 0.

Let $\mathcal{G}$ be any class of distribution functions in λ which contains G, and define

$$d_n = \inf_{\bar{G} \in \mathcal{G}} \rho(B_n, F_{\bar{G}}) \tag{16}$$

Let $G_n(\lambda) = G_n(x_1, x_2, \ldots, x_n; \lambda)$ be any element of $\mathcal{G}$ such that

$$\rho(B_n, F_n) \leq d_n + \varepsilon_n \tag{17}$$

Definition 2

We say that the sequence $G_n(\lambda) = G_n(x_1, \ldots, x_n; \lambda)$ of the random distribution functions in λ is effective for G if

$P[\lim_{n \to \infty} G_n(\lambda) = G(\lambda)$ at every continuity point of λ of $G] = 1$

for G in $\mathcal{G}$ (18)

Now we state a theorem.

Theorem 2

Assume that:

(i) For every fixed x, $F_\lambda(x)$ is a continuous function of λ.
(ii) The limits $F_{-\infty}(x) = \lim_{\lambda \to -\infty} F_\lambda(x)$ and $F_\infty(x) = \lim_{\lambda \to \infty} F_\lambda(x)$ exist for every x.
(iii) Neither $F_{-\infty}$ nor F_∞ is a distribution function of λ.
(iv) If G_1 and G_2 are any two distribution functions in λ such that $F_{G_1} = F_{G_2}$, then $G_1 = G_2$.

Then the sequence G_n defined by (17) satisfies (18) for the class $\mathcal{G}$ of all distribution functions in λ.

For the proof, see Robbins [30].

Theorem 3

(i) Assume all the hypotheses of Theorem 2, and let

$$\Lambda = \{0 \leq \lambda < \infty\}$$

instead of the whole real line.
(ii) Let the limit $F_\infty(x) = \lim_{\lambda \to \infty} F_\lambda(x)$ exist for every x.
(iii) F_∞ is not a distribution function.

(iv) If G_1, G_2 are any two distribution functions in λ which assign unit probability to $\Lambda = \{0 \leq \lambda < \infty\}$ such that

$$\int_\Lambda F_\lambda(x) \, dG_1(\lambda) = \int_\Lambda F_\lambda(x) \, dG_2(\lambda) \qquad \text{for all } x$$

then $G_1 = G_2$.

Then the sequence G_n defined by (17) satisfies (18) for the class $\mathcal{G}$ of all distributions which assign unit probability to Λ.

We shall illustrate this theorem by an example.

EXAMPLE

Define for $\lambda \in \Lambda = \{0 \leq \lambda < \infty\}$

$$F_\lambda(x) = \begin{cases} 0 & \text{for } x \leq 0 \\ \dfrac{x}{\lambda} & \text{for } 0 < x < \lambda \\ 1 & \text{for } x \geq \lambda \end{cases} \qquad (19)$$

Then

$$\lim_{\lambda \to 0} F_\lambda(x) = \begin{cases} 0 & \text{for } x \leq 0 \\ 1 & \text{for } x > 0 \end{cases} \qquad (20)$$

and

$$\lim_{\lambda \to \infty} F_\lambda(x) = 0 \qquad (21)$$

are not distributions, and (i), (ii) and (iii) hold.

For any G which assign unit probability to Λ we have, for $x > 0$,

$$F_G(x) = \int_\Lambda F_\lambda(x) \, dG(\lambda) = \int_{\{0 \leq \lambda \leq x\}} dG(\lambda) + x \int_{\{\lambda > x\}} \frac{dG(\lambda)}{\lambda} \qquad (22)$$

Hence if $F_{G_1} = F_{G_2}$, then

$$G_1(x) + x \int_{\{\lambda > x\}} \frac{dG_1(\lambda)}{\lambda} = G_2(x) + x \int_{\{\lambda > x\}} \frac{dG_2(\lambda)}{\lambda} \qquad (23)$$

If x is any common continuity point of G_1 and G_2, then

$$\int_{\{\lambda > x\}} \frac{dG_j(\lambda)}{\lambda} = \left[\frac{G_j(\lambda)}{\lambda} \right]_x^\infty + \int_{\{\lambda > x\}} \frac{G_j(\lambda)}{\lambda^2} \, d\lambda$$

$$= -\frac{G_j(x)}{x} + \int_{\{\lambda > x\}} \frac{G_j(\lambda)}{\lambda^2} \, d\lambda \qquad (24)$$

so that

$$G_j(x) + x \int_{\{\lambda > x\}} \frac{dG_j(x)}{\lambda} = x \int_{\{\lambda > x\}} \frac{G_j(\lambda)}{\lambda^2} \, d\lambda \qquad (25)$$

and hence from (23),

$$\int_{\{\lambda>x\}} \frac{G_1(\lambda)}{\lambda^2} d\lambda = \int_{\{\lambda>x\}} \frac{G_2(\lambda)}{\lambda^2} d\lambda \qquad (26)$$

at every continuity point $x > 0$ of G_1 and G_2. Differentiating with respect to x gives $G_1(x) = G_2(x)$ for every such x, and hence $G_1 = G_2$. One can refer to [45] for the problem of identifiability, i.e., under what conditions assumption (iv) holds.

4 ☐ A CASE IN WHICH ASYMPTOTIC OPTIMALITY DOES NOT EXIST BUT ALL IS NOT LOST

Let X be a random variable with density function

$$P(X = 1) = \lambda \qquad \text{where } \Lambda = \{0 \leq \lambda \leq 1\}$$
$$P(X = 0) = 1 - \lambda$$

On the basis of a single observation of X we want to estimate λ; if our estimate is $a \in A = \{0 \leq a \leq 1\}$, the loss will be taken to be $L(a,\lambda) = (\lambda - a)^2$. A density function t is determined by the two constants $t(0), t(1)$ which are available to us on the unit interval A; the risk in using t for a given λ is

$$R(t,\lambda) = (1-\lambda)[\lambda - t(0)]^2 + \lambda[\lambda - t(1)]^2$$
$$= t^2(0) + [t^2(1) - 2t(0) - t^2(0)]\lambda + [1 - 2t(1) + 2t(0)]\lambda^2 \qquad (1)$$

Consider the particular family of df's t_α defined for $0 < \alpha < 1$ by setting

$$t_\alpha(0) = \tfrac{1}{2}\alpha, \qquad t_\alpha(1) = \tfrac{1}{2}(1+\lambda) \qquad (2)$$

It can be seen from (1) that

$$R(t_\alpha,\lambda) = \tfrac{1}{4}[\alpha^2 + (1-2\alpha)\lambda] \qquad (3)$$

For $\alpha = \tfrac{1}{2}$ denoting t_α by t^*, we have

$$t^*(0) = \tfrac{1}{4} \qquad t^*(1) = \tfrac{3}{4} \qquad R(t^*,\lambda) = \tfrac{1}{16} \qquad (4)$$

For any a priori distribution G of λ let

$$\nu_i = \nu_i(G) = \int_0^1 \lambda^i \, dG(\lambda) \qquad i = 1, 2 \qquad (5)$$

Then from (1) it follows that for any density function t
$$R(t,G) = \int_0^1 R(t,\lambda) \, dG(\lambda)$$
$$= t^2(0) + \nu_1[t^2(1) - 2t(0) - t^2(0)] + \nu_2[1 - 2t(1) + 2t(0)] \quad (6)$$

without considering the cases in which $\nu_1 = 0$ or 1 we have after some simple algebra the formula

$$R(t,G) = \frac{(\nu_1 - \nu_2)(\nu_2 - \nu_1{}^2)}{\nu_1(1 - \nu_1)} + (1 - \nu_1)\left[t(0) - \frac{\nu_1 - \nu_2}{1 - \nu_1}\right]^2$$
$$+ \nu_1\left[t(1) - \frac{\nu_2}{\nu_1}\right]^2 \quad (7)$$

By the Bayes density function t_α for which

$$t_G(0) = \frac{\nu_1 - \nu_2}{1 - \nu_1} \qquad t_\alpha(1) = \frac{\nu_2}{\nu_1} \quad (8)$$

with

$$R(G) = R(t_G,G) = \frac{(\nu_1 - \nu_2)(\nu_2 - \nu_1{}^2)}{\nu_1(1 - \nu_1)} \quad (9)$$

$R(t,G)$ is minimized for a given G. Each t_α (and in particular t^*) is a Bayes density function; it is enough to find a distribution of λ such that

$$\frac{\nu_1 - \nu_2}{1 - \nu_1} = \frac{1}{2}\alpha \qquad \frac{\nu_2}{\nu_1} = \frac{1 + \alpha}{\alpha} \quad (10)$$

which is provided by the density

$$[\beta(\alpha, 1 - \alpha)]^{-1}\lambda^{\alpha-1}(1 - \lambda)^{(1-\alpha)-1} \quad (11)$$

for which

$$\nu_1 \equiv \alpha \qquad \nu_2 = \tfrac{1}{2}\alpha(1 + \alpha) \qquad R(G_\alpha) = \tfrac{1}{4}\alpha(1 - \alpha) \quad (12)$$

The fact that t^* is the Bayes density function relative to $G_{\frac{1}{2}}$ and since

$$R(t^*,G) = \int_0^1 R(t^*,\lambda) \, dG(\lambda) = \tfrac{1}{16} \quad (13)$$

$$\sup_G R(t,G) > \tfrac{1}{16} \qquad \text{for every } t \neq t^* \quad (14)$$

For if for some t', $\sup_G R(t',G) \leq \tfrac{1}{16}$, then in particular

$$\tfrac{1}{16} = R(t^*,G_{\frac{1}{2}}) \leq R(t',G_{\frac{1}{2}}) \leq \tfrac{1}{16} \quad (15)$$

so that

$$R(t',G_{\frac{1}{2}}) = R(t^*,G_{\frac{1}{2}}) = \tfrac{1}{16} \quad (16)$$

and therefore $t' = t^*$. Thus t^* is the unique "minimax" density function in the sense that it minimizes the maximum Bayes risk relative to the class of all a priori distributions G. Since G is unknown, it is therefore not unreasonable to use t^*; the Bayes risk will then be $\frac{1}{16}$ irrespective of G, while for any $t \neq t^*$ the Bayes risk will be $> \frac{1}{16}$ for some G (in particular for any G with $\nu_1 = \frac{1}{2}$, $\nu_2 = \frac{3}{8}$, for example, $G_{\frac{1}{2}}$).

For any $0 < \alpha < 1$ let $\mathcal{G}_\alpha$ denote the class of all G such that $\nu_1(G) = \alpha$, $0 < \alpha < 1$. For any G in $\mathcal{G}_\alpha$ (in particular for G_α) we see from (2) and (7) after some calculations that

$$R(t_\alpha, G) = \tfrac{1}{4}\alpha(1 - \alpha) \qquad (G \in \mathcal{G}_\alpha) \tag{17}$$

whatever may be the value of $\nu_2(G)$. It therefore follows as above that

$$\sup_{G \in \mathcal{G}_\alpha} R(t, G) > \tfrac{1}{4}\alpha(1 - \alpha) \qquad \text{for every } t \neq t_\alpha \tag{18}$$

so that relative to the class $\mathcal{G}_\alpha$, t_α is the unique minimax density function in the sense that it minimizes the maximum Bayes risk relative to the class $\mathcal{G}_\alpha$. When it is only known that $\nu_1(G) = \alpha$, it is therefore not unreasonable to use t_α; the Bayes risk will then be $\tfrac{1}{4}\alpha(1 - \alpha)$, while for any other density function the Bayes risk will be $> \tfrac{1}{4}\alpha(1 - \alpha)$ for some G in $\mathcal{G}_\alpha$ [in particular for any G with $\nu_1 = \alpha$, $\nu_2 = \tfrac{1}{2}\alpha(1 + \alpha)$, for example, $\mathcal{G}_\alpha$].

It follows from the above that

$$\frac{(\nu_1 - \nu_2)(\nu_2 - \nu_1^2)}{\nu_1(1 - \nu_1)} \leq \frac{1}{4}\nu_1(1 - \nu_1) \leq \tfrac{1}{16} \tag{19}$$

equality holding, respectively, when $\nu_2 = \tfrac{1}{2}\nu_1(1 + \nu_1)$ and when $\nu_1 = \tfrac{1}{2}$.

Suppose now that with unknown $\mathcal{G}$ we face this estimation problem repeatedly. The sequence $x_1, x_2, \ldots$ is an independent and identically distributed sequence of 0s and 1s with

$$P(x_i = 1) = \int_0^1 \lambda \, dG(\lambda) = \nu_1(G) \qquad P(x_i = 0) = 1 - \nu_1(G) \tag{20}$$

Hence the distribution of the X_i depends only on $\nu_1(G)$. Since t_G, defined by (8), involves $\nu_2(G)$ also, it follows that no $\varphi(x)$ can be asymptotically optimal relative to every G in a class $\mathcal{G}$ unless ν_2 is a function of ν_1 in $\mathcal{G}$, which is not likely to be the case in practice.

On the other hand, suppose

$$u_n = \frac{1}{n} \sum_{i=1}^n X_i \tag{21}$$

and consider the estimate $\tilde{T} = \{t_n\}$ with

$$t_n(0) = \tfrac{1}{2} u_n \qquad t_n(1) = \tfrac{1}{2}(1 + u_n) \tag{22}$$

For any G in $\mathcal{G}_\alpha$, by the law of large numbers, $u_n \to \alpha$, and hence $t_n \to t_\alpha$ with probability 1 as $n \to \infty$. Since we have

$$E(X_i) = \alpha = E(X_i^2) \qquad \text{var } X_i = \alpha(1 - \alpha) \tag{23}$$

it follows that

$$E(u_n) = \alpha \qquad E(u_n^2) = \text{var } u_n + \alpha^2 = \frac{\alpha(1-\alpha)}{n} + \alpha^2 \tag{24}$$

and hence from (6) it follows that

$$R_n(\tilde{T},G) = E\{\tfrac{1}{4}u_n^2 + \alpha[\tfrac{1}{4}(1 + 2u_n + u_n^2) - u_n - \tfrac{1}{4}u_n^2]$$
$$+ \nu_2(1 - 1 - u_n + u_n)\}$$
$$= \tfrac{1}{4}E(u_n^2 - 2\alpha u_n + \alpha) = \tfrac{1}{4}\alpha(1 - \alpha)\left(1 + \frac{1}{n}\right)$$
$$= R(t_\alpha,G)\left(1 + \frac{1}{n}\right) \tag{25}$$

Thus for large n we shall do almost as well by using $\tilde{T}$ as we could if we knew $\nu_1(G) = \alpha$ and used t_α. We have in fact for $G \in \mathcal{G}_\alpha$,

$$R_n(\tilde{T},G) - R(t_\alpha,G) = \frac{\alpha(1-\alpha)}{4n} \leq \frac{1}{16n} \tag{26}$$

while

$$R_n(\tilde{T},G) - R(t^*,G) = \frac{\alpha(1-\alpha)(n+1)}{4n} - \frac{1}{16}$$
$$= -[\tfrac{1}{4}(1 - 2\alpha)]^2 + \frac{\alpha(1-\alpha)}{4n} \tag{27}$$

Thus it illustrates the fact that even when an asymptotically optimal T does not exist or when it does exist but $R_n(T,G)$ is too slowly converging to $R(G)$, it may be worthwhile to use a T.

5 □ SMOOTH EMPIRICAL BAYES ESTIMATION

In a series of papers Maritz [16–18] developed a method of estimating the prior distribution and then using this estimate to give empirical Bayes estimates. We shall consider only continuous conditional density functions; an analogous method is available for discrete conditional density functions.

Initially Maritz approximates the prior distribution G with a step function G_k of k equal steps. The points of increase are at $\alpha_1, \alpha_2, \ldots, \alpha_k$.

The past experience $\langle \mathbf{x}_n \rangle = \{x_1, x_2 \ldots, x_n\}$, $n > k$, is then used to estimate G_k to give $G_{k,n}$. Finally we obtain

$$\delta_{k,n}(x) = \frac{\int \lambda f(x|\lambda) \, dG_{k,n}(\lambda)}{\int f(x|\lambda) \, dG_{k,n}(\lambda)} \qquad (1)$$

Hopefully, $\delta_{k,n}(x)$ is such that

$$\lim_{n \to \infty} R[\delta_{k,n}(X)] \leq R[\delta_G(X)] + \varepsilon_k$$

where ε_k is a constant dependent upon the number of points of increase of G_k.

We define

$$f_k(x) = \int f(x|\lambda) \, dG_k(\lambda)$$
$$f_G(x) = \int f(x|\lambda) \, dG(\lambda)$$

where the observation space $\mathfrak{X}$ is assumed to be the real line. For the observations $x_1, x_2, \ldots, x_n$ the approximate log likelihood is

$$\log L_k = \sum_{i=1}^{n} \log f_k(x_i)$$
$$= \sum_{i=1}^{n} \log \frac{1}{k-1} \sum_{j=1}^{k-1} \frac{1}{\lambda_{j+1} - \lambda_j} \int_{\lambda_j}^{\lambda_{j+1}} f(x_i|\lambda) \, d\lambda \qquad (2)$$

This is maximized with respect to $\lambda_1, \lambda_2, \ldots, \lambda_k$, and the maximizing values are denoted by $\alpha_1, \alpha_2, \ldots, \alpha_k$.

We define a quantity

$$D(G, G_k) = \int \log [f_k(x)] f_G(x) \, dx$$

Now

$$\frac{1}{n} \log L_k \to \int \log [f_k(x)] f_G(x) \, dx$$

in probability by Khintchine's theorem. Hence maximization of $\log L_k$ is equivalent to the maximization of $D(G, G_k)$. Now $D(G, G_k)$ is related to the measure of discrepancy between two density functions which was used by Kullback [15]

$$I(f_G, f_k) = -\int \log \frac{f_k(x)}{f_G(x)} f_G(x) \, dx$$
$$= \int \log [f_G(x)] f_G(x) \, dx - \int \log [f_k(x)] f_G(x) \, dx$$
$$= D - D(G, G_k)$$

Thus the determination of G_k by maximum-likelihood methods is equivalent to minimizing $I(f_G,f_k)$. Kullback [15, theorem 3.1, p. 14] shows that $I(f_G,f_k) \geq 0$ and is zero if and only if $f_G = f_k$. We want to show first that $I(f_G,f_k) \to 0$ as $k \to \infty$ (in practice $n > k$).

Let us define a special step function G_k^* of k equal steps at the points $\alpha_1^*, \alpha_2^* \ldots \alpha_k^*$ defined by

$$\alpha_r^* = G^{-1}\left(\frac{2r-1}{2k}\right) \qquad r = 1, 2, \ldots, k$$

As $k \to \infty$, $G_k^* \to G$ at every point of continuity of G; thus by the Helly-Bray lemma $f_k^*(x) \to f_G(x)$ for every x. This latter convergence is not necessarily uniform, but we shall prove that it is under certain conditions.

Suppose that the integrals $I(f_G,f_k^*)$ are uniformly bounded for all k and that $f(x|\lambda)$ is bounded for all $\lambda \in \Omega$. Using a theorem from Cramér [2, p. 67], we see that both $f_k^*(x)$ and $f_G(x)$ are continuous. For every x_0 in $\mathscr{X}$ we can find an open neighborhood $N_{x_0,k} \subset \mathscr{X}$ such that for all $x \in N_{x_0,k}$ we have that

$$1 - \delta(x_0,k) \leq \frac{f_k^*(x)}{f_G(x)} < 1 + \delta(x_0,k)$$

Define A_k as a closed set in $\mathscr{X}$ such that

$$\int_{A_k} f_G(x)\, dx \geq 1 - \varepsilon$$

for some small $\varepsilon > 0$. A_k is a compact set covered by the uncountable class of neighborhoods $N_{x,k}$. From this class we can extract a finite covering by the Heine-Borel property. From this finite covering find the maximum $\delta(x,k)$. Hence we have that $f_k^*(x)/f_G(x) \to 1$ uniformly for $x \in A_k$ as $k \to \infty$. Since $\int_{\mathscr{X}-A_k} f_G(x)\, dx < \varepsilon$, we get that

$$\frac{f_k^*(x)}{f_G(x)} \to 1 \qquad \text{uniformly in } x \text{ as } k \to \infty \qquad (3)$$

From a theorem of Kullback [15, p. 71] condition (3) is necessary and sufficient that

$$I(f_G, f_k^*) \to 0 \qquad \text{as } k \to \infty$$

Since $f_k(x)$ is determined by the G_k which minimizes $I(f_G, f_k)$ for fixed k, we get that
$$I(f_G, f_k^*) \geq I(f_G, f_k) \to 0 \quad \text{as } k \to \infty$$
Also we get that
$$f_k(x) \to f_G(x) \quad \text{uniformly in } x$$

To summarize the results so far: if we approximate G by a step function of k steps at the points determined by the maximization of $\log L_k$, Eq. (2), then $f_k(x)$ converges to $f_G(x)$.

What remains to be demonstrated is that $\delta_k(x) \to \delta_G(x)$ if $f_k(x) \to f_G(x)$. For simplicity of the argument we do not follow Maritz but make different assumptions. Let us assume that Ω is a bounded space. By the Helly-Bray lemma, since $f(x|\lambda)$ was earlier assumed to be bounded for all λ,
$$\int \lambda f(x|\lambda) \, dG_k(\lambda) \to \int \lambda f(x|\lambda) \, dG(\lambda)$$
Now
$$\delta_k(x) = \frac{\int \lambda f(x|\lambda) \, dG_k(\lambda)}{\int f(x|\lambda) \, dG_k(\lambda)}$$
and
$$\delta_k(x) \to \delta_G(x) \quad \text{for all } x \text{ as } k \to \infty$$

For fixed k the points of increase are the points $\alpha_1, \alpha_2, \ldots, \alpha_k$. For a sample of size n these points are estimated by $\hat{\alpha}_{1,n}, \hat{\alpha}_{2,n}, \ldots, \hat{\alpha}_{k,n}$. From the conditions for the consistency of maximum-likelihood estimators given in Chap. 9, Sec. 3, Theorem 1, we get that $\hat{\alpha}_{j,n} \to \alpha_j$ in probability. Hence $G_{k,n} \to G_k$ as $n \to \infty$ for fixed k, where $G_{k,n}$ is the step function of equal steps at the points $\hat{\alpha}_{1,n}, \ldots, \hat{\alpha}_{k,n}$. Thus we have proved that
$$\delta_{k,n}(x) \to \delta_G(x)$$

Concluding remarks

The empirical Bayes method of estimation is in a rapid state of growth. The usefulness of empirical Bayes techniques also has been found in hypothesis testing, decision theory, compound decision problems, classification problems, etc. The references for this chapter constitute an exhaustive bibliography on the subject, which may be useful to researchers interested in empirical Bayes techniques.

238   PARAMETRIC ESTIMATION

PROBLEMS

1. (Rutherford and Krutchkoff) Let X be a random variable with density function satisfying the relation

$$\frac{P_\lambda(x+1) - P_\lambda(x)}{P_\lambda(x)} = a(x) + b(x)\lambda \quad \text{for each } \lambda \in \Lambda$$

where $a(x)$ and $b(x)$ are any function of x, but not of λ, and $b(x) \neq 0$. Then show that

$$E_n(\lambda|x) = \frac{1}{b(x)}\left[\frac{P_n(x+1)}{P_n(x)} - 1\right] - \frac{a(x)}{b(x)}$$

is a consistent sequence of estimates of $E(\lambda|x)$. State the densities for which the assumption of this form holds.

2. Let X be a continuous random variable with density function $f_\lambda(x)$ which satisfies

$$\frac{f'_\lambda(x)}{f_\lambda(x)} = c(x) + d(x)\lambda \quad \text{for each } \lambda \in \Lambda$$

where $c(x)$ and $d(x)$ are functions of x, but not of λ, and $d(x) \neq 0$. Show that

$$E_n(\lambda|x) = \frac{1}{d(x)}\frac{f'_n(x)}{f_n(x)} - \frac{c(x)}{d(x)} \quad \text{for each } x \in R_1$$

is a consistent sequence of estimates of $E(\lambda|x)$. State the densities which can fall in this category. *Hint:* Use the following as estimator of $f(x)$

$$f_n(x) = \frac{1}{2\pi n^{\frac{1}{2}}} \sum_{i=1}^{n} \left(\frac{\sin y_i}{y_i}\right)^2$$

where

$$y_i = \frac{x - x_i}{2n^{-\frac{1}{2}}}$$

and

$$f'_n(x) = n^{\frac{1}{2}}[f_n(x + n^{-\frac{1}{2}}) - f_n(x)]$$

3. Let

$$F_0(x) = \begin{cases} 0 & \text{for } x < 0 \\ 1 & \text{for } x > 0 \end{cases}$$

and for $0 < \lambda < \infty$ let

$$F_\lambda(x) = \sum_{0 \le i \le x} e^{-\lambda} \frac{\lambda^i}{i!}$$

Prove that conditions (i) to (iv) of Theorem 3 of Sec. 3 hold.

4. Prove the following statement: An estimator based upon a sufficient statistic has the same risk as one based upon the original observations. State the precise conditions under which it is true.

5. (Rutherford) Find the form of the $E(\lambda|x)$ for the following density functions:

(a) $f(x|\lambda) = \dfrac{1}{\sqrt{2\pi}} e^{-(x-\lambda)^2/2}$ $\begin{array}{l} -\infty < \lambda < \infty \\ -\infty < x < \infty \end{array}$

(b) $p(x|\lambda) = \dbinom{n}{x} \lambda^x (1-\lambda)^{n-x}$ $\begin{array}{l} x = 0, 1, 2, \ldots, n \\ 0 \le \lambda \le 1 \end{array}$

(c) $f(x|\lambda) = \lambda e^{-x\lambda}$ $\begin{array}{l} \infty > \lambda > 0 \\ \infty > x > 0 \end{array}$

(d) $f(x|\lambda) = \dfrac{1}{\lambda} e^{-x/\lambda}$ $\begin{array}{l} \infty > \lambda > 0 \\ \infty > x > 0 \end{array}$

(e) $p(x|\lambda) = \dfrac{-1}{\log(1-\lambda)} \dfrac{\lambda^x}{x}$ $\begin{array}{l} x = 1, 2, \ldots \\ 0 \le \lambda \le 1 \end{array}$

6. Let $x_1, x_2, \ldots, x_n$ be independent random variables with common density function $N(\lambda, \sigma_x^2)$, and let λ be a random variable with density function $N(\mu, \sigma_\lambda^2)$.

(a) Find the marginal density of $(x_1, \ldots, x_n)$.
(b) Find the conditional density of λ given $(x_1, \ldots, x_n)$.
(c) Prove that

$$E(\lambda|x_1, \ldots, x_n) = \frac{\mu/\sigma_\lambda^2 + n\bar{x}/\sigma_x^2}{1/\sigma_\lambda^2 + n/\sigma_x^2}$$

where

$$\bar{x} = \frac{1}{n} \sum_1^n x_i$$

(d) Assume that we have a number m of observations on $(x_1, \ldots, x_n)$, where m is fairly large. What will be a reasonable estimate of $E(\lambda|x_1, \ldots, x_n)$?

7. Give an empirical Bayes estimation procedure for the parameter of the following density:

$$P(X = x) = \begin{cases} \lambda(1-\lambda)^{x-1} & x = 1, 2, \ldots, \infty \\ & 0 < \lambda > 1 \\ 0 & \text{otherwise} \end{cases}$$

8. (Rutherford and Krutchkoff) There is a sequence of realizations (x_1,λ_1), (x_2,λ_2), ..., (x_n,λ_n) each element being distributed according to the joint density function $f(x|\lambda)g(\lambda)$. The x's are observable; the λ's are not observable. The prior density function $g(\lambda)$ is an unspecified member of the Pearson family of curves; the conditional density function $f(x|\lambda)$ is known, and there must be known functions $h_j(x)$ such that $E[h_j(x)|\lambda] = \lambda^j$, $j = 1, 2, 3, 4$.

(a) Prove that if $E(\Lambda^4) < \infty$ and if

$$\hat{\mu}'_{j,n} = \frac{1}{n}\sum_{i=1}^{n} h_j(x_i)$$

then

$$\hat{\mu}'_{j,n} \xrightarrow{\text{a.s.}} E(\Lambda^j)$$

(b) Find the functions $h_j(x)$ for the Poisson density; normal density with unknown mean and known variance σ^2; binomial density. For what other density function do the functions $h_j(x)$ exist? Not exist?

9. (Rutherford and Krutchkoff) It is known that the Pearson curves are continuous functions of their first four moments with certain minor restrictions (see Dumas [3]). To denote the dependence of the prior density upon its moments we write $g(\lambda) = g(\lambda;\mathbf{\mu})$ where $\mathbf{\mu} = (\mu_1, \mu_2, \mu_3, \mu_4)$. Let $g_n(x) = g(\lambda',\mathbf{\mu}_n)$, where $\mathbf{\mu}_n = (\hat{\mu}_{1,n}, \hat{\mu}_{2,n}, \hat{\mu}_{3,n}, \hat{\mu}_{4,n})$, denote the estimate of the prior density based upon the estimates of the central moments derived from the estimated row moments given in (a) of Prob. 8.

(a) Prove

$$g_n(\lambda) \xrightarrow{\text{a.s.}} g(\lambda)$$

Hint: See Pratt [25].

(b) Prove that

$$\lim_{h \to \infty} E\{[G_n(\lambda) - G(\lambda)]^2\} = 0 \quad \text{a.e. } \lambda$$

where
$$G(\lambda) = \int_{-\infty}^{\lambda} g(v)\, dv$$
$$G_n(\lambda) = \int_{-\infty}^{\lambda} g_n(v)\, dv$$

10. (Rutherford) The posterior distribution of Λ given x is defined by

$$P(\Lambda \leq \lambda | x) = \frac{\int_{-\infty}^{\lambda} f(x|v)\, dG(v)}{f_G(x)} \tag{1}$$

where
$$f_G(x) = \int_{-\infty}^{\infty} f(x|\lambda)\, dG(\lambda)$$

Let $P_n(\Lambda \leq \lambda | x)$ denote the estimate of the posterior distribution determined from Eq. (1) by substituting $G_n(\lambda)$ for $G(\lambda)$. Prove that

$$P_n(\Lambda \leq \lambda | x) \xrightarrow{a.s.} P(\Lambda \leq \lambda | x) \qquad \text{a.e. } \lambda$$
$$\text{a.e. } x$$

11. (Rutherford and Krutchkoff) In bayesian point estimation of the realization of $\Lambda\ (= \lambda)$ with absolute difference as loss function the median M_x of the posterior distribution of Λ given x is the Bayes estimate. Define an estimated median $M_{x,n}$ by

$$P_n(\Lambda \leq M_{x,n} | x) = 0.5$$

using the result of Prob. 3.

(a) Prove that the Bayes estimate is the posterior median.

(b) Prove that
$$\lim_{n \to \infty} E(|M_{x,n} - M_x|) = 0 \qquad \text{a.e. } x$$

i.e., that $M_{x,n}$ is an asymptotically optimal empirical Bayes point estimate.

12. (Parzen) Assume the setup of Sec. 2 and define

$$f_n(x) = \frac{1}{nh} \sum_{j=1}^{n} W\left(\frac{x - X_j}{h}\right)$$

where h and W are functions satisfying the following conditions:

(i) $\lim_{n \to \infty} h(n) = 0$.

(ii) $\lim_{n \to \infty} nh(n) = \infty$.

(iii) W is an even Borel measurable function.
(iv) $\sup_{-\infty < y < \infty} |W(y)| < \infty$.
(v) $\int |W(y)|\, dy < \infty$.
(vi) $\lim_{y \to \infty} |yW(y)| = 0$.
(vii) $\int W(y)\, dy = 1$.

Then prove that
$$\lim_{n \to \infty} E[|f_n(x) - f_G(x)|^2] = 0$$

at all points of continuity of $f_G(x)$, and give examples of W. *Hint:* A simple example of a weighting function is

$$W(y) = \begin{cases} \tfrac{1}{2} & |y| \leq 1 \\ 0 & |y| > 1 \end{cases}$$

and

$$f_n(x) = \frac{1}{nh} \sum_{i=1}^n W\left(\frac{x - X_j}{h}\right)$$
$$= \frac{F_n(x + h) - F_n(x - h)}{2}$$

where

$$F_n(x) = \frac{1}{n}\{\text{number of indices } j \text{ such that } X_j \leq x\}$$

Another weighting function is

$$W(y) = \frac{1}{2\pi}\left[\frac{\sin(y/2)}{y/2}\right]^2$$

13. Prove or disprove that the empirical Bayes procedure defined in Sec. 2 is inadmissible.

14. For δ_{kn} defined in Sec. 5, state conditions under which

$$\lim_{n \to \infty} \lim_{k \to \infty} R[\delta_{k,n}(x)] = R[\delta_G(x)]$$

and prove the result.

REFERENCES

1 CACOULTOS, T.: Estimation of a Multivariate Density, *Am. Inst. Math.*, vol. 18, pp. 179–189, 1966.

2 CRAMÉR, H.: "Mathematical Methods of Statistics," Princeton University Press, Princeton, N.J., 1946.

3 DUMAS, M.: Sur les courbes de fréquence de K. Pearson, *Biometrika*, vol. 35, pp. 113–117, 1948.

4 HALMOS, P. R.: "Measure Theory," D. Van Nostrand Company, Inc., Princeton, N.J., 1950.

5 HANNAN, J. F.: Approximation to Bayes Risk in Repeated Play, *Contr. Theory Games*, vol. 3, pp. 97–139, 1956.

6 HANNAN, J. F., and H. ROBBINS: Asymptotic Solutions of the Compound Decision Problem for Two Completely Specified Distributions, *Ann. Math. Statist.*, vol. 26, pp. 37–51, 1955.

7 HANNAN, J. F., and J. R. VAN RYZIN: Rate of Convergence in the Compound Decision Problem for Two Completely Specified Distributions, *Ann. Math. Statist.*, vol. 36, pp. 1743–1752, 1965.

8 JOHNS, M. V., JR.: Non-parametric Empirical Bayes Procedures, *Ann. Math. Statist.*, vol. 28, pp. 649–669, 1957.

9 JOHNS, M. V., JR.: An Empirical Bayes Approach to Non-parametric Two-way Classifications, in H. Solomon (ed.), "Studies in Item Analysis and Prediction," pp. 221–232, Stanford University Press, Stanford, Calif., 1961.

10 JOHNS, M. V., JR.: Two-action Compound Decision Problems, *Stanford Univ. Dept. Statist. Tech. Rept.* 87, 1966.

11 JOHNSON, N. L., E. NIXON, D. E. AMOS, and E. S. PEARSON: Tables of Percentage Points of Pearson Curves for B_1 and B_2 Expressed in Standard Measure, *Biometrika*, vol. 50, pp. 459–498, 1963.

12 KAGAN, A. M.: An Empirical Bayes Approach to the Estimation Problem, *Soviet Math.*, vol. 3, pp. 1760–1762, 1962.

13 KAGAN, A. M.: Note on the Robbins Scheme, *Soviet Math.*, vol. 4, pp. 739–742, 1963.

14 KRUTCHKOFF, R. G.: A Supplementary Sample Non-parametric Empirical Bayes Approach to Some Statistical Decision Problems, *Biometrika*, vol. 54, pp. 451–458, 1967.

15 KULLBACK, S.: "Information Theory and Statistics," John Wiley & Sons, Inc., New York, 1959.

16 MARITZ, J. S.: Smooth Empirical Bayes Estimation for One-parameter Discrete Distributions, *Biometrika*, vol. 53, pp. 417–429, 1967.

17 MARITZ, J. S.: On the Smooth Empirical Bayes Approach to Testing of Hypothesis and the Compound Decision Problem, *Biometrika*, vol. 55, pp. 83–100, 1968.

18 MARITZ, J. S.: Smooth Empirical Bayes Estimation for Continuous Distribution, *Biometrika*, vol. 54, pp. 435–450, 1967.

19 MARTZ, H. F.: Empirical Bayes Estimation in Multiple Linear Regression, Ph.D. thesis, Virginia Polytechnic Institute, 1967.

20 MIYASAWA, K.: An Empirical Bayes Estimation of the Mean of a Normal Population, *Bull. Intern. Statist. Inst.*, vol. 38, no. 4, pp. 181–188, 1960.

21 MOSTELLER, F., and D. L. WALLACE: Inference in an Authorship Problem, *J. Am. Statist. Assoc.*, vol. 58, pp. 275–309, 1963.

22 NEYMAN, J.: Two Breakthroughs in the Theory of Statistical Decision Making, *Rev. Inst. Intern. Statist.*, vol. 30, pp. 11–27, 1962.

23 PARZEN, E.: On Estimation of a Probability Density Function and Mode, *Ann. Math. Statist.*, vol. 33, pp. 1065–1076, 1962.

24 PEARSON, E. S.: Some Problems Arising in Approximating to Probability Distributions, Using Moments, *Biometrika*, vol. 50, pp. 95–112, 1963.

25 PRATT, J. W.: On a General Concept of "In Probability," *Ann. Math. Statist.*, vol. 30, pp. 549–558, 1959.

26 RAIFFA, H., and R. SCHLAIFER: Applied Statistical Decision Theory, Harvard University Press, Cambridge, Mass., 1961.

27 ROBBINS, H.: Asymptotically Subminimax Solutions of Compound Statistical Decision Problems, *Proc. 2d Berkeley Symp. Math. Statist. Probability*, pp. 131–148, 1951.

28 ROBBINS, H.: An Empirical Bayes Approach to Statistics, *Proc. 3d Berkeley Symp. Math. Statist. Probability*, pp. 157–163, 1955.

29 ROBBINS, H.: The Empirical Bayes Approach to Testing Statistical Hypotheses, *Rev. Inst. Intern. Statist.*, vol. 31, pp. 195–208, 1963.

30 ROBBINS, H.: The Empirical Bayes Approach to Statistical Decision Problems, *Ann. Math. Statist.*, vol. 35, pp. 1–20, 1964.

31 ROY, L. K., and M. T. WASAN: The First Passage Time Distribution of Brownian Motion of Positive Drift, *Math. Biosci.*, vol. 2, pp. 191–204, 1968.

32 ROY, L. K., and M. T. WASAN: Properties of the Time Distribution of Standard Brownian Motion, *Rev. Trabajos Estastist.*, vol. 19, pp. 67–82, 1968.

33 RUTHERFORD, J. R.: Some Parametric Empirical Bayes Techniques, Ph.D. thesis, Virginia Polytechnic Institute, 1966.

34 RUTHERFORD, J. R.: The Empirical Bayes Approach: Estimating Posterior Quantiles, *Biometrika*, vol. 54, pp. 672–675, 1967.

35 RUTHERFORD, J. R., and R. G. KRUTCHKOFF: The Empirical Bayes Approach: Estimating the Prior Distribution, *Biometrika*, vol. 54, pp. 326–328, 1967.

36 RUTHERFORD, J. R., and R. G. KRUTCHKOFF: ε-asymptotic optimality of empirical Bayes estimator, *Biometrika*, vol. 56, pp. 220–223, 1969.

37 SAMUEL, E.: An Empirical Bayes Approach to the Testing of Certain Parametric Hypotheses, *Ann. Math. Statist.*, vol. 34, pp. 1370–1385, 1963.

38 SAMUEL, E.: Convergence of the Losses of Certain Decision Rules for the Sequential Compound Decision Problem, *Ann. Math. Statist.*, vol. 35, pp. 1606–1621, 1964.

39 SAMUEL, E.: Sequential Compound Estimators, *Ann. Math. Statist.*, vol. 36, pp. 878–889, 1965.

40 SAMUEL, E.: On Simple Rules for the Compound Decision Problem, *J. Roy. Statist. Soc.*, ser. B, vol. 27, pp. 238–244, 1965.

41 SAMUEL, E.: The Compound Decision Problem in the Opponent Case, *Israel J. Math.*, vol. 3, pp. 117–126, 1965.

42 SAMUEL, E.: Sequential Compound Rules for the Finite Decision Problem, *J. Roy. Statist. Soc.*, ser. B, vol. 28, pp. 63–72, 1966.

43 Scheffé, H.: A Useful Convergence Theorem for Probability Distributions, *Ann. Math. Statist.*, vol. 18, pp. 434–438, 1947.

44 Swain, D. D.: Bounds and Rates of Convergence for the Extended Compound Estimation Problem in the Sequence Case, *Stanford Univ. Dept. Statist. Tech. Rept.* 81, 1965.

45 Teicher, H.: Identifiability of Mixtures, *Ann. Math. Statist.*, vol. 32, pp. 244–248, 1961.

46 Van Ryzin, J. R.: The Compound Decision Problem with $m \times n$ Finite Loss Matrix, *Ann. Math. Statist.*, vol. 37, pp. 412–424, 1966.

47 Van Ryzin, J. R.: The Sequential Compound Decision Problem with $m \times n$ Finite Loss Matrix, *Ann. Math. Statist.*, vol. 37, pp. 854–975, 1966.

48 Van Ryzin, J. R.: Repetitive Play in Finite Statistical Games with Unknown Distribution, *Ann. Math. Statist.*, vol. 37, pp. 976–994, 1966.

49 Van Ryzin, J. R., and M. V. Johns, Jr.: Convergence Rates for Empirical Bayes Two-action Problem I, Discrete Case, *Stanford Univ. Dept. Statist. Tech. Rept.* 131, 1967.

50 Van Ryzin, J. R., and M. V. Johns, Jr.: Convergence Rates for Empirical Bayes Two-action Problems II, Continuous Case, *Stanford Univ. Dept. Statist. Tech. Rept.* 132, 1967.

51 Von Mises, R.: On the Correct Use of Bayes Formula, *Ann. Math. Statist.*, vol. 13, pp. 156–165, 1942.

52 Wilks, S. S.: "Mathematical Statistics," John Wiley & Sons, Inc., New York, 1962.

Index

Index

Absolute error, 104, 206
Absolutely continuous, 135
Accessible points, 192
Adaptive estimation, 226
Additive number theory, 105
Admissibility, 11, 182–184, 207
Admissible, 114, 182, 183, 190, 191, 194–198, 213, 215, 218
Admissible estimate, 183, 192, 216
Aitken, A. C., 148
Ali, Mir M., 29, 34, 35
Amos, D. E., 243
Analytic function, 65, 73
Analytic integrals, 65
Antle, C. E., 50, 53, 55
A posteriori, 220
Approximation theory, 1, 36

A priori, 11, 14, 184, 185, 187, 190, 192, 194–197, 202, 204–206, 212, 219, 231, 233, 240
Arens, B. E., 55
As good as, 183
Asymptotic distribution, 29
Asymptotic properties, 7
Asymptotic relative efficiency, 31, 34
Asymptotic theory, 32
Asymptotically best linear unbiased estimate, 30
Asymptotically efficient estimate, 158
  in strict sense, 154, 173
Asymptotically modal unbiased estimate, 125
Asymptotically normal, 158, 161

249

Asymptotically optimal, 226, 227, 231, 234, 241
Asymptotically optimal procedure, 218

Bahadur, R. R., 74, 75, 94, 95
Barankin, E. W., 75, 79, 95, 110, 132, 148
Barnard, G., 15
Barton, D. E., 175, 178
Basu, A. P., 163, 178
Bayes density, 232
Bayes estimate, 182, 184, 185, 189, 192, 195–197, 202, 204–206, 220
Bayes estimator, 184, 187
Bayes function, 221
Bayes method, 182
Bayes risk, 189, 205, 222, 233
Bayes solution, 189
Bayesian, 218, 241
Bellman, R., 15
Best asymptotically normal, 14, 125
Best linear unbiased estimate, 31
Best unbiased linear combination, 131
Better than, 183, 184, 210
Bhattacharyya, A., 144, 148
Bhattacharyya bounds, 134, 144
Biased estimate, 9, 152, 205
Bijective function, 166
Birnbaum, A., 105, 132
Blackwell, D., 15, 102, 111, 112, 115, 117, 132, 216
Blyth, C. R., 216
Bosewell, M. T., 178
Boundary point, 192
Bounded unbiased estimate, 68
Boundedly complete, 57, 68, 69, 72
Boundedly completeness, 57
Breiman, L., 16
Brownian motion, 225
Brunk, H. D., 150, 178
Burkholder, D. L., 75, 95

Cacoultos, T., 242
Cauchy distribution, 63
Chapman, D. G., 134, 149
Chapman-Robbins-Kiefer inequality, 142
Chapman-Robbins-Kiefer lower bound, 147
Chernoff, H., 15
Classification problem, 237
Closed, 192
Complete, 57–60, 62, 64, 65, 66, 68, 69, 71, 72, 104, 106, 111, 112, 115, 116, 166, 169
Complete family of density functions, 3, 4
Complete family of estimates, 184, 215
Completeness, 56, 57, 64, 65, 198
Compound decision problems, 237
Condition of regularity, 134
Conditional risk, 112
Confidence interval, 13
Confidence relation, 163, 164
Confidence-set method, 12
Consistency, 4, 14
  of maximum-likelihood estimators, 237
Consistent, 154, 158, 161
Consistent estimate, 4, 173, 224, 225
Consistent sequence of estimates, 4, 5, 224–226, 238
Constant expected loss, 196
Constant risk, 182, 184, 195, 197
Convex, 97, 101, 102, 114, 193
Convex function, 96–98, 102
Convexity, 97, 99
Correlated case, 24
Countably additive, 56
Cramér, H., 15, 132, 136, 148, 178, 243
Cramér-Rao inequality, 134, 135, 144, 146, 182, 197, 214
Cramér-Rao lower bound, 138, 139, 147, 157, 158, 161
Cramér-Rao result, 139
Criterion of goodness of estimates, 118

Crossover, 177
Curtailed symmetric rule, 15

David, F. N., 105
David, H. A., 55
Davis, D. J., 178
Decision, 2, 14, 218, 237
Decision theory, 129
DeGroot, M. H., 15
De la Garza, A., 36, 37, 55
Dempster, A. P., 14, 16
Density-unbiased estimate, 132
Deutsch, R., 16, 162, 178
Dimension, 29, 31, 57, 79
Distance criteria, 129
Dominated case, 74, 75, 81
Downtown, F., 35
Dumas, M., 240, 243

Eeden, C. van, 151, 178, 179
Efficiency, 125
Elfving, G., 55
Empirical approach, 218
Empirical Bayes, 14, 218, 226, 237, 240, 241
Empirical Bayes estimate, 223
Empirical Bayes procedure, 242
Empirical distribution, 220
Empirical estimation, 226
Empirical frequency, 220
Empirical procedure, 223
ϵ-good strategy, 193
Epstein, B., 174, 179
Equivalent, 183, 215
Essentially complete family of estimates, 215
Essentially equivalent, 117
Estimable, 21–23
Estimable parameter, 115
Estimation, 14, 218
Estimation problem, 96
Estimation theory, 1, 2, 150, 157

Euler notation, 29
Expected loss, 12, 96, 126, 193

Factorization criterion, 77
Factorization theorem, 83
Failure time, 162
Fend, A. V., 134, 149
Fiducial probability, 13
Fisher, R. A., 6, 16, 74, 150, 179
Fixed-sample procedure, 15
Flehinger, B. J., 162, 179
Fraser, D. A. S., 14, 16, 73, 75, 86, 95, 132, 134, 149, 179
Functionally independent, 148

Game theory, 10, 182, 193, 194
Gart, J. J., 147–149
Gauss, C. F., 6, 16, 105, 150
Gauss-Markoff theorem, 18, 22, 24, 46
General order statistic, 87, 114
Generalization, 144
Generalized density, 94
Generalized least squares, 26, 30
Generalized probability measure, 89
Generalized variance, 25, 29, 30, 46–48, 54
Girshick, M. A., 15, 16, 102, 132, 203, 216
Good strategy, 194
Greenberg, B. G., 29, 35
Group, 12
Group of transformations, 207
Grouped samples, 152
Guest, P. G., 36, 50, 55
Guttman, I., 134, 149

Hader, R. J., 176, 180
Halmos, P. R., 73, 74, 81–83, 95, 228, 243
Halperin, M., 161, 179
Hannan, J. F., 243

Hardy, G., 102
Hazard function, 163
Heine-Borel property, 236
Helly-Bray lemma, 237
Higher dimension, 79
Hilderbrand, F. B., 52, 55
Hodges, J. L., Jr., 139, 184, 197, 212, 216
Hoel, P. G., 36, 53–55
Holder's inequality, 101
Homogeneous polynomial, 66
Homomorphic, 12
Householder, A. S., 162, 179
Hunt, G., 11
Huzurbazar, V. S., 161, 179

Inaccessible points, 192
Inadmissibility, 184
Inadmissible, 242
Index, 192
Inferential techniques, 2, 3
Information, 37, 39, 40, 45, 74
Information matrix, 178
Interval estimation, 12
Invariance, 207
Invariance principle, 11, 182
Invariant, 207–209
Invariant estimate, 11, 12, 208, 210
Invariant minimax estimate, 208
Inverse function theorem, 80
Iterated integral, 70

Jensen's inequality, 100, 112
Johns, M. V., Jr., 243
Johnson, N. L., 243
Joint asymptotic efficiency, 34, 35

Kagan, A. M., 243
Kale, B. R., 162, 179
Karlin, S., 124, 216
Katz M., Jr., 95
Katz, M. W., 213, 216

Kendall, M. G., 16, 55, 157, 179
Khintchine's theorem, 159, 235
Kiefer, J., 37, 55, 134, 149
Kolmogorov, A. N., 131, 132
Koopman, B. O., 75, 79, 95
Krutchkoff, R. G., 238, 240, 241, 243
Kudo, H., 216
Kullback, S., 235, 236, 243
Kulldorff, G., 29, 33, 35, 152, 153, 156, 173, 179

Lagrange multiplier, 48
Lagrangian coefficient, 51
Lagrangian interpolation, 37
Laplace transform, 60, 63
Large sample, 14, 29, 150, 157, 158, 161
Laurent, A. G., 163, 179
Least favourable, 192, 194
Least squares, 18, 32, 33, 150
Least-squares estimates, 6
Least-squares estimators, 19, 20, 22, 23, 27, 45
Least-squares fit, 152
Least-squares principle, 6
Lebesgue, 63, 64
Lebesgue bounded-convergence theorem, 228
Lebesgue dominated-convergence theorem, 228
Lebesgue measure, 57, 67, 90, 91, 137, 169
LeCam, L., 5, 16, 179
Legendre differential equation, 50
Legendre polynomial, 48, 50, 52
Lehmann, E. L., 16, 64, 70, 72–74, 92, 94, 95, 115, 132, 139, 179, 184, 197, 212, 216
Levine, A., 36, 53, 54
Likelihood equation, 154, 157, 161, 172
Likelihood function, 151, 153, 166, 167
Lindeberg-Levy theorem, 160
Lindgren, B. W., 89, 95
Linear estimate, 136

Linear unbiased estimator, 24
Littlewood, J., 102
Lloyd, E. H., 32, 33, 35
Lobatto quadrature, 52
Local conditional sufficiency, 75
Locally best unbiased estimates, 104, 109, 110
Location and scale parameters, 18, 26
Locations, 39
Loeve, M., 180
Loss function, 2, 3, 10, 96, 105, 109, 114, 115, 126, 132, 185, 186, 189, 197, 205, 207, 212–214, 219
Lower bound of variance, 134, 140, 146

Maitra, A. P., 79, 95
Maritz, J. S., 218, 234, 243
Markoff, A., 105
Markoff estimators, 24–26, 46
Marshall, A. W., 180
Martz, H. F., 243
Maxima of variance, 51
Maximum-likelihood estimate, 8, 14, 31, 34, 152–154, 161, 166–170, 172, 173, 175, 177, 214
Maximum-likelihood estimation, 150–152, 171
Maximum-likelihood method, 7, 150, 151
Maximum variance, 45, 53
Mean-unbiased estimate, 9, 111, 125–128, 132
Mean-unbiased estimation, 104, 150
Measure of discrepancy, 235
Median, 118, 119
Median-unbiased estimate, 104, 119, 128, 132
Median unbiasedness, 118
Mendenhall, W., 176, 180
Method of covariance, 117
Method of maximum likelihood, 7, 150, 151
Method of moments, 5

Method of transforms, 104
Minima of variance, 52
Minimal dimension, 75
Minimal sufficiency, 91
Minimal sufficient statistics, 74, 79, 89, 91, 92
Minimax, 114
Minimax density function, 232, 233
Minimax estimate, 10, 182, 183, 186, 192, 193, 195–198, 208, 212–214, 216, 218
Minimax estimation, 182, 184, 191
Minimax estimator, 183, 186, 188–190
Minimax invariant estimate, 215
Minimax procedure, 184
Minimax variance, 36, 51
Minimaxity, 10, 15, 182, 207
Minimum-chi-square estimate, 7
Minimum-chi-square method, 7, 14
Minimum-mean-square-error estimate, 9
Minimum risk, 128
Minimum-risk estimators, 117
Minimum variance, 25, 27, 31, 117, 157, 162, 169, 170, 172, 174, 175
Minimum-variance estimate, 108
Minimum-variance-unbiased estimate, 9, 26, 111, 114, 115, 117, 118, 130, 134
Minimum-variance-unbiased linear estimators, 24, 46
Mission time, 162
Mitra, S., 131, 132
Miyasawa, K., 243
Modal-unbiased estimate, 104, 120–127, 132
Modal unbiasedness, 120
Mode, 120, 157
Modeling, 162
Moment problem, 36
Monotone function, 61, 151
Monotone parameters, 151
Mosteller, F., 15, 16, 132, 243

Newton-Raphson process, 162
Neyman, J., 7, 12, 14, 16, 74, 77, 95, 105, 244
Neyman criterion, 78, 93
Nixon, E., 243
Nonbayesian, 218
Noncompleteness, 63
Nonparametric, 226
Nonrandomized, 191, 197, 214
Nonrandomized estimate, 113
Nonrandomized unbiased estimator, 114
Normal equations, 6, 19, 20, 45
Nuisance parameters, 139, 140
Null set, 56

Ogawa, J., 29, 35
Optimal locations, 50
   of regressor variables, 46
Optimal spacing of regressor variables, 18, 36
Optimality criterion, 218, 226
Optimum design, 37
Optimum order statistics, 29
Optimum spacing, 31, 34, 45
Optimum stopping rules, 134
Order statistic, 66, 67, 86, 87, 93, 128, 163
Orthogonal projection, 20

Parametric function, 21
Partially grouped sample, 173
Parzen, E., 241, 244
Patil, G. P., 105, 132
Payoff, 193
Pearson, E. S., 243, 244
Pearson family of curves, 240
Pearson type of distribution, 5
Percentile, 129
Pitcher, T. S., 95
Pitman, E., 130
Point estimate, 2, 3

Point-estimation method, 12
Polya, G., 102
Polynomial, 59, 61, 62, 193
Polynomial regression, 54
Polynomial regression coefficients, 18
Polynomial regression model, 19
Pratt, J. W., 240, 244
Prevost, G., 50, 55
Principle of invariance, 11, 182
Principle of minimaxity, 10, 15, 182, 207
Principle of point estimation, 2, 3, 12
Principle of unbiasedness, 8–10, 170
Product probability measure, 86
Programming, 15
Projection, 22
Proschan, F., 180
Pugh, E. L., 163, 180
Pure strategy, 193, 194

Quadratic equation, 186
Quadratic function, 6, 204
Quadratic loss function, 202, 214
Quantiles, 29–31, 34, 35

Radon-Nikodym derivative, 84
Radon-Nikodym theorem, 68
Raiffa, H., 244
Randomized estimates, 113, 208
Randomized unbiased estimator, 114
Rao, C. R., 16, 111, 112, 115, 117, 132, 149, 180
Rao-Blackwell theorem, 111, 112, 115, 117, 169, 170
Regression coefficients, 24
Regression parameters, 46
Regressor variable, 36, 39, 50
Regularity conditions, 135, 136, 140, 146, 147, 153
Reliability, 162, 168, 169, 171
Reliability inference, 150, 162
Risk, 104, 105, 126, 183, 186, 187, 189, 205, 214, 219, 227, 231, 239

Risk function, 2, 11, 182–184, 198
Robbins, H., 14–16, 134, 218, 220, 226, 229, 244
Roy, L. K., 224, 226, 244
Roy, T., 131, 132
Rutherford, J. R., 238–241, 244

Saleh, A. K., 29, 34, 35
Samuel, E., 244
Sarhan, A. E., 29, 34, 35
Savage, L. J., 15, 16, 74, 81–83, 132, 203, 216
Scheffé, H., 16, 18, 35, 55, 70, 72–74, 92, 94, 115, 132, 245
Schlaifer, R., 244
Schur, I., 48, 55
Schwartz's inequality, 204
Sequential analysis, 15
Sequential estimation, 14, 15
Sequential estimation procedures, 134
Sequential procedure, 191, 192
Seth, G. R., 149
Shohat, J. A., 36, 55
Silverstone, H., 148
Slutky theorem, 161
Small sample, 182
Smaller dimension, 78
Smith, K., 36, 53, 55
Smooth empirical Bayes estimation, 234
Sobel, M., 179
Spacing, 36, 37, 39, 40
    for minimax variance, 50
Sprott, D. A., 14, 17
Squared-error loss function, 105, 128
Standard Brownian motion, 223
Stieltjes, T. J., 48
Stochastic approximation, 15
Stopping rule, 191
Strategy, 193
Strict convexity, 116
Strictly convex, 97, 98, 100, 101, 111, 115–117
Strong completeness, 57, 69, 71

Stronger property, 69
Strongly complete, 57, 69, 70, 72, 73
Stuart, A., 16, 55, 157
Subfields, 74
Sufficiency, 3, 74, 80, 81, 86, 150
Sufficient, 75–78, 81, 83, 85, 92, 166, 169
Sufficient statistics, 3, 74–79, 81, 84–86, 89, 92–94, 104, 111, 112, 115, 116, 148, 157, 164, 172, 239
Swain, D. D., 245
Symmetric, 88
Symmetric estimators, 114
Symmetric function, 67, 87

Tamarkin, J. D., 36, 55
Tate, R. F., 132, 163, 176, 180
Tchebychev points, 54
Tchebychev's inequality, 4
Teicher, H., 245
Testing hypotheses, 12
Theory of estimation, 1, 2, 150, 157
Theory of games, 10, 182, 193, 194
Thompson, W. A., Jr., 180
Two-persons-zero-sum game, 193

Unbiased, 171, 183, 205, 206
Unbiased estimate, 4, 10, 22, 25, 104–110, 116, 117, 129–131, 140, 142, 144, 146, 147, 157, 162, 169, 170, 174
    may be absurd, 108
    may not exist, 109
Unbiased estimation, 104, 105, 118, 182
Unbiased estimator, 111, 115, 118
Unbiased linear estimator, 24, 25
Unbiasedness, 8–10, 170
Uncorrelated observations, 37
Undominated case, 75
Unicity property, 63

Uniformly minimum risk, 112, 116
Uniformly minimum-risk function, 183
Uniformly minimum-risk invariant, 210, 212
Uniformly minimum variance, 104, 105
Uniformly minimum-variance estimate, 214
Uniformly minimum-variance-unbiased estimate, 9, 10, 105, 106, 127, 129, 131, 144, 215
Uniformly smaller, 126
Uniformly within interval, 66
Unique unbiased linear estimator, 22, 23
Uniqueness theorem, 65, 165

Vandermonde determinant, 47
Vandermonde form matrix, 38, 40
Van Ryzin, J. R., 245
Von Mises, R., 245
Von Neumann, J., 10

Wald, A., 10, 14, 15, 17, 132, 216
Wallace, D. L., 243
Wasan, M. T., 15, 17, 132, 212, 213, 216, 224, 226, 244
Wilks, S. S., 180, 245
Wolfowitz, J., 15, 17, 27, 55, 137, 140, 149, 216
Wolfowitz regularity conditions, 135

QA
3
Q38
#9
c.1

DEC 14 1973